Surveillance, Privacy and Public Space

Today, public space has become a fruitful venue for surveillance of many kinds. Emerging surveillance technologies used by governments, corporations, and even individual members of the public are reshaping the very nature of physical public space. Especially in urban environments, the ability of individuals to remain private or anonymous is being challenged.

Surveillance, Privacy, and Public Space problematizes our traditional understanding of "public space". The chapter authors explore intertwined concepts to develop current privacy theory and frame future scholarly debate on the regulation of surveillance in public spaces. This book also explores alternative understandings of the impacts that modern living and technological progress have on the experience of being in public, as well as the very nature of what public space really is.

Representing a range of disciplines and methods, this book provides a broad overview of the changing nature of public space and the complex interactions between emerging forms of surveillance and personal privacy in these public spaces. It will appeal to scholars and students in a variety of academic disciplines, including sociology, surveillance studies, urban studies, philosophy, law, communication and media studies, political science, and criminology.

Bryce Clayton Newell is an Assistant Professor in the School of Information Science at the University of Kentucky. In his research, he focuses on understanding the impact that surveillance and information and communication technologies (ICTs) have on individuals, society, and the law.

Tjerk Timan is a policy advisor and researcher at TNO, the Netherlands. He has been publishing on topics of policing technologies, surveillance theory and practices, and privacy. Recently, he has co-edited a book on privacy in public space.

Bert-Jaap Koops is Professor of Regulation & Technology at the Tilburg Institute for Law, Technology, and Society (TILT), Tilburg University. In 2016/17, he was Distinguished Lorentz Fellow at the Netherlands Institute for Advanced Study (NIAS). He publishes widely on cybercrime, cyber-investigation, privacy, and data protection, including recently "A Typology of Privacy Spaces" and "Bentham, Deleuze and Beyond."

Routledge Studies in Surveillance

Kirstie Ball, William Webster, Charles Raab

https://www.routledge.com/Routledge-Studies-in-Surveillance/book-series/RSSURV

1 Big Data, Surveillance and Crisis Management
Edited by Chiara Fonio and Kees Boersma

2 Surveillance, Privacy and Public Space
Edited by Bryce Clayton Newell, Tjerk Timan, and Bert-Jaap Koops

3 Surveillance and Democracy in Europe
Kirstie Ball and William Webster

Surveillance, Privacy and Public Space

Edited by
Bryce Clayton Newell, Tjerk Timan
and Bert-Jaap Koops

Routledge
Taylor & Francis Group
LONDON AND NEW YORK

First published 2018
by Routledge
2 Park Square, Milton Park, Abingdon, Oxon OX14 4RN

and by Routledge
711 Third Avenue, New York, NY 10017

Routledge is an imprint of the Taylor & Francis Group, an informa business

British Library Cataloguing in Publication Data
A catalogue record for this book is available from the British Library

Library of Congress Cataloging in Publication Data
A catalog record has been requested for this book

ISBN: 978-1-138-70996-6 (hbk)
ISBN: 9781315200811 (ebk)

Typeset in Bembo
by Taylor & Francis Books

Contents

Illustrations

Figures

Tables

Contributors

Melis Baş is a PhD candidate at the Philosophy Department of the University of Twente. She works on a project entitled, "Mediated Public Spaces: Parks and Politics." In this project, Melis interprets Hannah Arendt as a philosopher of technology, and situates public parks as political agents in our everyday practices of politics.

Anthony Boanada-Fuchs acts as project and academic relations manager at the University of St. Gallen Institute of Management in Latin America (GIMLA). He is also post-doctoral researcher at CASE (Centre for Research on Architecture, Society & the Built Environment) at ETH Zurich and has held other post-doc and visiting positions at the University of Lausanne, University Basel and the University of São Paulo. He completed his PhD at the Graduate Institute of International and Development Studies in Geneva (IHEID), Switzerland.

John Chuang is a Professor at the UC Berkeley School of Information, and Director of the BioSENSE Lab, with research interests in bio-sensory computing applications, security, and privacy.

Silvia de Conca is a PhD Researcher at the Tilburg Institute for Law, Technology, and Society (TILT) at Tilburg University, where she works at the intersection of robotics, artificial intelligence, and privacy. She holds an LL. M. in IT, Media & Communication Law from the London School of Economics and Political Science (London), and an LL.M. with Honors in International and Comparative Law from the University of Roma Tre (Rome). Before starting at TILT, Silvia lived in Mexico, where she was Assistant Professor of Information Technology Law and Intellectual Property Law at the Law Department of the Tec de Monterrey University ITESM (Chihuahua).

Stuart Hargreaves is an Assistant Professor at the Faculty of Law of the Chinese University of Hong Kong, where he is also Director of the LLB Programme and Assistant Dean for Undergraduate Studies. He is a barrister and solicitor (Law Society of Upper Canada). He holds a BA from McGill

University, a JD from Osgoode Hall Law School, a BCL from the University of Oxford, and an SJD from the University of Toronto.

Michael A. Katell is a PhD Candidate in data and technology ethics at the University of Washington Information School. His research interests include privacy, digital reputation, digital equity, and the ethical design of information systems.

Bert-Jaap Koops is Professor of Regulation & Technology at the Tilburg Institute for Law, Technology, and Society (TILT), the Netherlands. In 2016/17, he was Distinguished Lorentz Fellow at the Netherlands Institute for Advanced Study (NIAS). His main research fields are cybercrime, cyber-investigation, privacy, and data protection.

Primož Križnar earned a Bachelor of Laws in 2013 and Master of Laws in 2015. He is currently working as a senior judicial advisor at the District state prosecutor's office in Ljubljana and as a legal researcher at the Institute of Criminology at the Faculty of Law in Ljubljana.

Olya Kudina is a PhD candidate at the Philosophy Department of the University of Twente. She works on a project entitled, "Mediated Morality: Google Glass, Sex Selection, and the Complex Interactions between Ethics And Technology." Her research contributes to conceptualizing the way technologies co-shape human norms and values.

Bryce Clayton Newell is an Assistant Professor in the School of Information Science at the University of Kentucky. He is also a licensed attorney (California), a documentary filmmaker, and a former Google Policy Fellow. He specializes in the study of surveillance, information practices, and the use of ICTs in the contexts of policing and irregular migration, with additional research and teaching interests in criminal law, information policy, and technology regulation.

Gerard Jan Ritsema van Eck MA is a PhD researcher at the Security, Technology, and *e*-Privacy (ST*e*P) research group at the University of Groningen. His research focuses on the use of smartphones for participatory surveillance and its impact on the free enjoyment of human rights in public spaces.

Patrick Schmidt is Professor and Chair of Political Science at Macalester College. His teaching and research interests span American constitutional law, regulation, and public policy, including higher education and information policy.

Elaine Sedenberg is a PhD Candidate at the UC Berkeley School of Information and Co-Director of the Center for Technology, Society & Policy (CTSP). She studies information sharing practices and the legal, ethical, and economic aspects of data access for research purposes.

Sarah Shoemaker graduated from Macalester College with a Bachelor of Arts in Political Science in 2015. She subsequently received a Fulbright award to Malaysia.

Kristen Thomasen is an Assistant Professor of Law, Robotics and Society at the University of Windsor Faculty of Law. She is also a PhD Candidate in Law at the University of Ottawa. Her research explores the legal and social impacts of automated technology, particularly on privacy in public space. Kristen is a member of the Law Society of Alberta. Prior to starting her PhD, Kristen clerked for the Honourable Madam Rosalie Abella at the Supreme Court of Canada, and for the Alberta Court of Queen's Bench.

Tjerk Timan is a policy advisor and researcher at TNO in the field of privacy, surveillance, and data science. Previously, he held a position as postdoctoral researcher at the Tilburg Institute for Law, Technology, and Society (TILT), where he was engaged in questions around new privacy issues in the 21st century. He has also been a lecturer at the Erasmus University Rotterdam, providing courses on digital humanities, big data, and data visualization. He completed his PhD thesis at the University of Twente in which he investigated emerging and existing technologies of surveillance in urban nightlife districts.

Daniel Trottier (PhD, Queen's Canada) is an Associate Professor in the Department of Media and Communication at Erasmus University Rotterdam. His primary research interests include user-led forms of monitoring and policing through digital media.

Carissa Véliz is a Research Fellow at the Uehiro Centre for Practical Ethics at the University of Oxford. She completed her doctoral degree (DPhil) in Philosophy at the University of Oxford, writing her dissertation on the ethics and politics of privacy.

Richmond Wong is a PhD Candidate at the UC Berkeley School of Information. His research investigates how speculative and critical design methods can be used to interrogate privacy and other policy-related concerns during the process of technology design.

Aleš Završnik, Doctor of Law (LL.D.), is a Senior Research Fellow at the Institute of Criminology at the Faculty of Law in Ljubljana and Associate Professor at the Faculty of Law University of Ljubljana. Currently, he is EURIAS Research Fellow 2017-2018 at the Collegium Helveticum in Zürich, co-funded by Marie-Sklodowska-Curie Actions (COFUND Programme - FP7).

Acknowledgements

The preparation and editing of this book was made possible by a grant from the Netherlands Organisation for Scientific Research (NWO), project number 453-14-004.

Chapter 1

Privacy and surveillance in the streets

An introduction

Bryce Clayton Newell

Introduction

Privacy in public space is a recurring topic within a variety of academic fields. Even prior to Warren and Brandeis's (1890) famous call to protect privacy in the face of advancing methods of photography in 1890, new technologies and individual privacy have been in tension with each other. Today, emerging surveillance technologies used by governments, corporations, and even individual members of the public are reshaping the very nature of physical public space and are challenging the ability of individuals to remain private or anonymous in these spaces, especially in urban environments. Within these spaces, we see complex interactions arise between anonymity and identification; between trade/commercial activity, social interaction, and suspicion. Surveillance and privacy intersect in these spaces, and these intersections are significant, especially as information traditionally kept in private physical spaces or personal activities not subject to long-term and aggregate monitoring are recorded and captured by a variety of visual and other sensors embedded in public space.

Personal information and documents that once stayed (physically) protected behind closed doors or inside peoples' homes are now stored in the cloud (a so-called "digital home" (Irion 2015), albeit with much more porous walls), and this information is also transmitted through physical public space, subject to capture and analysis (consider, for example, the use of cell-tower simulators, or "stingrays", that mimic cellular towers and capture calls and data in transit over cellular networks). These developments have resulted in a metaphorical "evaporation" of the home (Koops 2014, 247, 256–257), traditionally seen as the bastion of privacy, as a barrier between surveillance and the details of private life. This "evaporation of houses" (Koops 2014, 257) and the increasing accessibility of private information in and from public space is connected to "ubiquitous trackability" because, as this private data spills over into public space, it adds to the digital profiles already being collected by surveillance in these spaces and, in turn, this aggregation makes further analysis possible that can reveal additional "aspects of private life that used to be confined to private places" (Koops

2014, 259). As such, public space has become a ripe and fruitful venue for surveillance of many kinds.

Privacy and surveillance may both implicate the control of information (Marx 2015a), although each may not be limited only to informational concerns (Koops et al. 2017). Surveillance has been influentially defined as "the focused, systematic and routine attention to personal details for purposes of influence, management, protection or direction" (Lyon 2007, 14). Haggerty and Ericson (2006, 3) also argue that surveillance "involves the collection and analysis of information about populations in order to govern their activities." The definition of privacy, on the other hand, has been highly contested for decades. Privacy has been defined in many ways, both normatively and descriptively (Koops et al. 2017, 492; Marx 2015a, 35; Moore 2010; Newell 2015, 28; Solove et al. 2006, 40–51). For purposes of this introduction, it is sufficient to say that privacy and surveillance may oppose one another in many contexts but, as Marx (2015a, 33) has argued, they should not necessarily be seen as opposites. In the end, it seems apparent that we need to find new boundary-marking concepts to more appropriately determine the scope of privacy rights for individuals or groups in public space and regulate the scope and techniques of surveillance directed at them. Complicating matters, of course, is the fact that privacy and expectations of privacy may differ by culture (see, for example, Friedman et al. 2008). Despite that concern, however, it is likely that some aspects of privacy are shared across cultures (Moore 2010, 49–52) and that certain boundary-marking concepts might be used, or be at least usable, in many different cultures (for example, "personal space"). If this is true, then the cross-cultural (and even time-dependent) variability lies in where the boundary itself is placed – for example, how and where exactly do we draw the line between "personal space" and "not personal space"?

In the chapters that follow, various authors identify ways in which privacy and surveillance interact and intersect in public space. These analyses provide insights, empirical findings, and arguments that can help illuminate possible boundary-marking concepts and frame future scholarly debate about how we should regulate (or not regulate) surveillance in public space. They also argue for alternative understandings of how we should understand the impacts that modern ways of living and technological progress have on the experience of being in public and – at the extreme – the very nature of what public space really is. The contributions are also highly multidisciplinary, and range in methods from philosophical to empirical to legal, and provide a broad overview of the changing nature of public space and the complex interactions between emerging forms of surveillance and personal privacy in these public spaces.

In the next section, I summarize how the authors of the subsequent chapters conceptualize and problematize the notion and definition of "public space", as this provides a useful lens through which to view the ensuing analyses of the role of privacy and surveillance in these spaces. Following this summary of what the "public space" is (or can be), I then outline a series of concepts proposed by the authors that are frequently related and often intertwined – but that present possible avenues for improving current privacy theory and that may serve as useful boundary-marking concepts for improving how we currently regulate surveillance (and balance the interests in both privacy and surveillance) in public spaces.

The nature of public space

A number of authors in this book argue that "the idea of public space is considerably larger than its mere spatial dimension" (Fuchs). Véliz identifies a spatial dimension supplemented by considerations of funding and ownership. Trottier discusses both "embodied spaces" and "mediated publics." Fuchs, while outlining the development and change in the public spaces of São Paulo, Brazil, claims that public space also has political, social, cultural, and legal dimensions. "From such an understanding," Fuchs argues, "public spaces may also be perceived as a mirror of society and its cities that embodies the balance between publicness and privacy, freedom and restrictions, surveillance and security." Finding inspiration in "the analytical categories of Hannah Arendt's 'public realm' (2013) and Jürgen Habermas' 'public sphere' (Habermas 1989), which describe the zone of life beyond the state itself where individuals create a collective identity," Shoemaker and Schmidt identify two dimensions of public space: a *physical* dimension and a *performative* dimension (in which individuals "perform public roles, such as the offices of public trust," or engage in other "forms of communal interaction where individuals address (in speech and action) matters of public concern"). Likewise, according to Fuchs, "The public manifests in moments when people lay down their differences and start interacting with one another, an argument central to Hannah Arendt's 'human condition,' but also the work of Jürgen Habermas on the public sphere (*Öffentlichkeit*)." Kudina and Baş also provide a definition of public space informed by Arendt's ideas, arguing that "public space is not necessarily a designated physical location," but rather is "a place for socialization, interaction, and representation of one's identity."

Similarly, public space should not be defined solely by ownership (Fuchs; Véliz; Shoemaker and Schmidt). To Véliz, neither the physical characteristics of a public place (e.g., a place accessible to others and where you will be visible or exposed) nor the matter of funding and ownership of places should be used as metrics to demarcate normative privacy rights. She argues that "privacy cannot be descriptively specified or normatively decided by reference to what is private in a financial sense" or what is private in the spatial sense. Fuchs makes the case that shopping malls exemplify how privately owned spaces can become "public" spaces. Shoemaker and Schmidt examine the university as a public space, "both the nature of its physical spaces and its ideational connection to the public realm." At their core, universities are public spaces, they argue, "because of how they present themselves to mass publics, but even more because they are sites of contest: inquiry, debate and a struggle for access, as individuals seek to use education and expertise as the pathway to control of the public realm." This implies that spaces might be public for physical/spatial reasons as well as for ideational reasons. Complicating matters, of course, is the possibility that universities, like many other "public" spaces, may also serve more private functions (and parts of these spaces may serve hybrid purposes).

Both Fuchs and Shoemaker and Schmidt show how fear and the desire for accountability have led to increased surveillance (in both São Paulo and on university campuses). Indeed, "[r]ecording devices, identification schemes, data collection and larger police budgets often appear as rational and relatively unintrusive responses, whatever the evidence of effectiveness" (Shoemaker and Schmidt). Surveillance of ideologically loaded public spaces (like universities) or activities in public space (like protests) provides the watchers with a greater ability to understand the attitudes, beliefs, and political views of those subject to such surveillance. Of course, achieving such law enforcement purposes often relies on the ability to accurately identify suspects. Indeed, identification has become crucial for the functioning of modern government. Interestingly, a number of authors also refer to surveillance (or surveillance technologies) as motivated by suspicion. Indeed, viewing surveillance as a means of capturing information about people in order to govern and influence their behavior, we might see surveillance technologies deployed as *technologies of suspicion*, and the data collected by them as subject to social constructions of suspicion, including human bias and potential for injustice and discrimination (as Katell posits; see also Trottier).

In some jurisdictions, privacy rights against state surveillance are limited (or virtually eliminated) based on the fact that such surveillance occurs in public spaces (see, for example, the discussion provided by Newell, de Conca, and Thomasen). Although these determinations likely owe their origins to a variety of factors, it is probably not coincidental that modern definitions of "public" often include the notion that what is public is "open to the scrutiny of anyone" (Sennett 1992, 16). Indeed, we find evidence of this in the ubiquity of surveillance cameras and other sensors in public spaces, as well as in commercial activities like the street-mapping projects of companies like Google, Microsoft, and Apple, that "digitally photograph or otherwise record elements of public space and then recreate them as 'explorable worlds'" (Hargreaves). Additionally, "[t]he transformation of public space into increasingly observed and mediated space follows from the idea that the right to privacy in public is an unreasonable expectation, particularly where surveillance is seen as means of fulfilling moral duties to protect and care for others" (Katell, citing Himma 2015; Gooas et al. 2015).

Concepts, distinctions, and the search for boundaries

The mosaic

A number of chapters throughout this book emerge from concerns regarding the growing "mosaic" of information about individuals collected and aggregated – either over time or *en masse* – through the operation of multiple forms of surveillance, rather than just the mere fact that surveillance is occurring in public space. The basic idea of the so-called *mosaic theory* is that the aggregation of numerous individual data points about a person can, in their composite, reveal substantially more about a person than any of the individual pieces of data can on their own – and, thus, also facilitate inferences about the person beyond what is evident in the collected data. Shoemaker and Schmidt identify the risks of the mosaic quite clearly in their chapter about the impact of surveillance on university campuses: "The potential of information systems to paint a nuanced portrait of a student over time undermines privacies that were once protected by the sheer inefficiency of collecting it all in one place." As summarized by Katell, "[s]cholars have been considering the transformative nature of digitally mediated, pervasive surveillance for decades as long-held perceptions of being anonymous in public have fallen away due to the relentless logging and analysis of our movements, communications, interactions, and transactions." The (potential) application of the mosaic theory to the regulation of police investigatory conduct in the United States has provoked quite a significant response from legal scholars in recent years (e.g., Gray and Citron 2013; Slobogin 2012), including a number of critical responses (e.g., Kerr 2012; Kugler and Strahilevitz 2015).

In chapter 6, Sedenberg, Wong, and Chuang examine the ability of sensors to detect biosignals ("heart rates, body temperatures, expressions, tones, involuntary reflexes, and biological signatures") from a distance or by analyzing biological samples (for example, hair or skin) left behind unwittingly by people as they traverse public spaces. They argue that, although "these signals have existed throughout human time," emerging biosensors allow those in control of the devices "to record continuously and *en masse*, and to systematically analyze our innate biological nature." Among other characteristics, these forms of "biosensed data" provide expansive amounts of information about individuals, especially when combined with data from different sources, potentially allowing "inferences" to be drawn "about the body … to deduce emotional states [and] physical reactions," possibilities that may even reveal information about "memories and thoughts." Thus, for Sedenberg, Wong, and Chuang, the *inferential power* granted to those wielding such aggregated information is what makes the surveillance capable of collecting, aggregating, and analyzing this information so problematic.

In their chapter, Kudina and Baş examine the ability of augmented reality technologies, such as Google Glass, to contribute to the persistent and pervasive nature of sensing and surveillance in public space. Augmented reality glasses, characterized by the well-known but ill-fated Google Glass project, were introduced into the public consciousness at a time when mass surveillance in public had already become a norm. However, as an example of an individual wearable technology, Google Glass brought practices of (and the potential for) observation to a new level. When wearables are watching, typically through small embedded sensors (e.g., cameras), they do so surreptitiously, or, at least, they create ambiguities and uncertainty about the extent and scope of observation taking place.

Later in the book, Završnik and Križnar examine the idea of the mosaic in the context of geolocation. Drawing from recent case law in the United States and the European Court of Human Rights, they echo the concerns of judges (and others) that, "disparate location data … one leaves behind in the digital age have grown to an unprecedented level. Seemingly irrelevant data on its own, such as phone logs or GPS data from smart phones, can reveal sensitive personal data on an aggregate level: where you live, where you work, when you travel, who you meet with, and who you sleep with." Završnik and Križnar offer an optimistic vision of the mosaic theory as a "new legal remedy within the European legal framework" with the potential to "further develop the legal standard of reasonable expectation of privacy in public space." They argue that society would benefit from legally recognizing that "disparate pieces of information may have no informative value on their own, but can have great value on an aggregate level" and that such recognition could "curtail disproportionate employment of mass surveillance technologies."

Relatedly, Katell argues that public space surveillance may also lead to "bias and discrimination against marginalized segments of society," in which "surveillance, open government records, reidentification of data subjects, and quantified reputation" all come together to disproportionally affect certain groups more than others. Surveillance, including forms advocated for by civil liberties advocates (e.g., body-worn police cameras), "is inherently biased" because of the way it is used and the purposes for which it is used (e.g., police work). The "overproduction of police data," or data created by other (non-police) surveillance in public spaces, also includes the capture of information about bystanders. These realities are connected to the realities of surveillance identified by Haggerty and Ericson (2000) (the "surveillant assemblage") and Lyon (2003; 2007) ("social sorting"), and the pervasiveness of surveillant sensors in public spaces "raise questions pertinent to public anonymity and privacy in general because they gather and retain information indiscriminately, sweeping up anyone and everyone who happens to be in sensor range of the device when it is active."

Voluntary versus involuntary emanation or disclosure

Many scholars engage with the popular concept of information privacy as the ability to control access to or uses of personal information (with its necessary consent element) (see, for example, Newell 2015; Moore 2010). However, drawing from concerns related to the mosaic described above, Sedenberg, Wong, and Chuang's work suggests that consent breaks down as a control mechanism in the face of surveillance that captures and aggregates personal information that is involuntarily leaked or emitted into the public sphere (for example, biosensors combining multiple types of biosensed data from public spaces). The same argument might apply to DNA traces left behind as people move in transit through public space. However, the boundaries of determining voluntariness are not clear. Véliz states that even seemingly voluntary exposure (for example, kissing in public) is sometimes a consequence of constrained choices, and thus not always completely voluntary with regards to place. However, she does not explicitly argue that such "voluntary" exposure ought to benefit from cognizable claims on others to act with discretion; after all, how would other persons know where the line is drawn in such circumstances? By calling out social norms, she contests attempts to delineate the private in universalistic terms.

Sedenberg, Wong, and Chuang argue that "developing and emerging technologies enable even more data streams to be collected remotely, through casual contact – meaning sensors that may be briefly, and perhaps unknowingly, touched through everyday interaction – or through physical traces left behind. This potential makes biosensed data collection with no opportunity for consent a new challenge for individuals in public spaces." Ritsema van Eck also notes that the photographs captured by people on the street can also capture biometric information of others, as well as potentially other personal information and metadata related to (for example, geo-location coordinates), without the others knowing. Additionally, Sedenberg, Wong, and Chuang describe how such technologically enabled surveillance can magnify, or augment, involuntarily leaked biological information detectable without such interventions, or provide inferences about a person's mental state of which the person is not even aware. Although, for example, "tears are thought to be a subtle cue to instigate empathy from others within a few feet," the potential to infer "biosensed data like emotions or internal physiological states" through systematic collection and analysis, "all signals become magnified beyond their original natural public scope." As such, it is also likely that this magnification will only be amplified as artificial intelligence and machine-learning technologies become more sophisticated than they already are.

One answer to this critique, of course, is that such information is either not private or that any expectation that such information is private is not objectively reasonable because the information is detectable within the public space without any physical intrusion into a person's body or private property (see, for example, Justice Binnie's argument in the Canadian Supreme Court decision of *R. v. Tessling*[1] that the heat emanating from the external walls of a house is reasonably subject to surveillance and that measurement of that heat by the police without a warrant, at least insofar as it does not "permit any inferences about the precise activity giving rise to the heat," is permissible)[2]. However, given the sensitivity of the personal information detectable through, for example, sophisticated biosensors that can operate at a distance, we might be concerned that such answers discount a fuller account of what privacy might (or ought to) mean. As only one of many possible alternatives, the collection of such information can be seen to implicate the potential for state or corporate domination of the individual, especially if these actors could use such information arbitrarily or capriciously to discriminate, deny benefits, or otherwise affect the individual's options, opportunities, or choices (see Newell 2015; Roberts 2015), reminiscent of the concerns expressed by Katell in his chapter. On the other hand, as Hargreaves notes, it can also be very difficult to regulate (especially private action; for example, public street photography) in this regard, "without also rendering actionable a whole host of innocuous behavior or running into (in most jurisdictions) constitutionally protected rights of expression." Hargreaves' critique is most relevant in contexts where the right of expression is engaged and, in those cases, involuntary emanation appears to be less useful as a distinction in jurisdictions with stronger commitments to freedom of expression. Of course, many forms of surveillance that potentially capture involuntary emanations will not necessarily implicate rights of expression or speech but, even in these cases, it seems questionable whether the involuntariness of the emanation (something that has to do with the *lack of consent*) is really more helpful than focusing, for example, on subsequent use of collected information or the potential for additional inferences to be drawn about the subject (a concern related to the mosaic theory, discussed above).

Practical obscurity versus theoretical visibility

Privacy law often ties privacy rights to the existence of a reasonable (legitimate) expectation of privacy (see Newell, de Conça, and Thomasen). In applying this test, courts in multiple jurisdictions have tied reasonableness (legitimacy) to theoretical visibility – that is, a person who *could*, in theory, be seen by another lawfully present person (e.g., a passerby), even if they were not actually seen by such a person, cannot claim a reasonable expectation of privacy because their actions were theoretically exposed to the public. Thus, theoretical visibility produces normative, negative implications for privacy in public space. In contrast, the notion of "practical obscurity" (Hartzog and Stutzman 2013a; 2013b; Selinger and Hartzog 2014) provides one interesting and potentially very useful way to push back on the normative power that theoretical visibility has thus far acquired. Practical obscurity (which, like the mosaic theory, emerged in U.S. case law involving the government's insistence that it should not have to disclose records under the federal Freedom of Information Act because doing so would allow the recipients to infer other, classified, information)[3], "focuses on identifying the 'fundamental ways information can be obtained or kept out of reach, correctly interpreted or misunderstood'" (Newell 2015, quoting Selinger and Hartzog 2014, 1). According to Hartzog and Stutzman (2013b, 48), this analytic notion "could serve as a metric for the boundary of allowable disclosure by information recipients" and might take the shape of "some form of mandated obscurity." In essence, the theory of practical obscurity bears a strong resemblance to aspects of Cohen's (2013; 2012) theory of "semantic discontinuity" – defined as "gaps and inconsistencies within systems of meaning, and to a resulting interstitial complexity that leaves room for the play of everyday practice" (Cohen 2012, 224; see also Newell 2015, 32–36).

In chapter 7, Katell shows how surveillance can operate to limit or defeat obscurity by explaining how some forms of mobile surveillance technology (for example, body-worn cameras) can observe and cast their gaze into "obscure locations" beyond the reach of static or stationary devices. Likewise, Sedenberg, Wong, and Chuang examine how a variety of sensors, deployed in publically accessible places, can uncover seemingly obscure and previously undetectable biological information about people. Véliz flips the table a little, arguing that "public" spaces may, at times, actually offer greater amounts of privacy – or something resembling practical obscurity – than even the most private spaces in the home. "The home," writes Véliz, "is not always the ideal place to find privacy." And when that becomes true, "the only escape is to go out, to the coffee shop, the public square." It is in these public, crowded spaces, she argues, that we can sometimes find protection not just for our personal information, but also for our *autotopos*, or "a metaphorical sensorial personal space or privacy bubble" that surrounds us.

Technological progress also facilitates privacy, pseudonymity, and anonymity. As various anti-surveillance technologies provide individuals with some level of practical obscurity, these same technologies may increase the cost of government (and corporate) surveillance, disrupting traditional social controls. If this is true, it provides all the more reason for privacy advocates to build obscurity-preserving measures into privacy law or other regulation of surveillance in public space. But what might these measures look like? Cohen's (2013, 2012) concept of "semantic discontinuity" suggests that "gaps in enforcement and in systems of surveillance and control" (Balkin 2012, 81) are ultimately desirable and that we should promote "institutional and technical structures" that meet these ends (Cohen 2012, 6). Ultimately, this means institutionalizing "disorganization" (Miller 2014, 134) into systems of control and surveillance, including the criminal justice system and the laws that regulate surveillance by both public and private actors. Framed differently, this "disorganization" can also be cast as maintaining levels of inefficiency that existed in the past (see, for example, Justice Alito's and Sotomayor's concurring opinions in *United States v. Jones*)[4] – as historically focused crippling of modern surveillance – that stands as a reminder that theories of legal protection have not kept pace with technological development. This view has even been adopted by Justices of the Supreme Court of the United States, who have argued, in the context of GPS-enabled location tracking, that the Court "must 'assur[e] preservation of that degree of privacy against government that existed when the Fourth Amendment was adopted,'"[5] and that modern surveillance that overcomes previous limitations of "police resources and community hostility" should be looked at with suspicion.[6] These arguments have strong emotional force, especially for those concerned about government surveillance, but we should also consider what introducing significant regulatory "gaps and inconsistencies" into the capabilities of law enforcement during criminal investigations might mean in terms of the broader societal goals of effective law enforcement, security, and criminal justice. Technologically driven realizations of practical obscurity may provide needed protections for individuals who are not already subjects of reasonable police suspicion, but they may also decrease the effectiveness of law enforcement more broadly, even after reasonable suspicion develops.

Civil inattention, social etiquette, and the virtue of discretion

A number of authors in this book argue, in various ways, that civil inattention, or the idea that people should exercise some level of discretion as a form of social etiquette, could provide the basis for a social norm that also promotes some aspects of practical obscurity and help avoid the negative (privacy-related) harms facilitated by the mosaic. According to Véliz, we should be able to expect others "to act with discretion" – essentially, to exercise appropriate civil inattention (Goffman 1972, 385) – in public places, at least in certain "sensitive" circumstances where our exposure is involuntary (Goffman, of course, also emphasized the importance of *showing* discretion). In a sense, this is a call for social norms to take the place of law (or more formal regulation) in incorporating obscurity into how individuals observe and watch each other in shared spaces. Kudina and Baş explain how Google's prompt to Glass users to "use [their] best judgment" about when to record exemplifies the desire for people to consciously perform social etiquette while, at the same time, the technology itself overcomes the ability to understand who is watching (or recording) because "there are no clear signals to communicate to the public whether the device is recording at any particular moment." Other forms of surveillance act similarly, institutionalizing a degree of ambiguity that may well hinder the development of norms of civil inattention.

Relatedly, Kudina and Baş also note arguments from Tonkiss (2003) that "an ethics of indifference [is] a necessary condition for coexisting in the public space" (Kudina and Baş), an idea Tonkiss (2003, 300) develops from the claim that "dissociation is, in fact, the basic form of urban sociation, one that allows us to coexist with all these largely unknown others. Not interacting with others in this sense becomes a primary condition for urban social life, securing individual calm together with relative social peace." In his chapter, Fuchs builds on Sennett's (1992, 27) claim that "silence in public became the only way one could experience public life" to argue that this need spurred people to increasingly engage in selective self-representation in public – essentially as a means to ensure some forms of dissociation (and civil inattention) or, at least, to redirect attention to only particular attributes that they preferred be made public.

However, how do we encode civil inattention into electronic systems of surveillance, mostly information and communication technologies (ICTs), and, from a normative perspective, do we want to? Is such encoding even possible, or would the increasing ubiquity of sensors (fixed or moving) present in our public space environments mean that what we really might achieve is some form of *technologically mediated attention* (with an attendant lack of ability to not pay attention in some circumstances). Véliz argues that social clues ought to inform our decorum – and civil inattention – but does not clearly describe how, or whether, law ought to be present in such circumstances (although she notes that this question is an important one for future research). However desirable encoding civil inattention into systems of surveillance might seem as a counter-move to the "new surveillance" (Marx 2003) and as an effort to regain practical obscurity, it runs head-on into the growing cry from civil society organizations and many members of the public to require more "objective" evidence collection and documentation by the police (for example, the call for police to deploy body-worn cameras (*Harvard Law Review* 2015, 1800; see also Ready and Young 2014)). Additionally, hard-coding certain forms of civil inattention into surveillance systems might also mean closing down opportunities for technological development – for example, the incorporation of augmented reality and biometric identification technologies into devices like Google Glass that call attention to identity and other aspects of the world (and people) around the user in (near) real-time.

Surveillance and participation

As surveillance has become more ubiquitous, it has also become more democratized (although panoptic theories of surveillance may still have significance in certain contexts) as the technologies of surveillance proliferate throughout society and, for example, can fit into trouser pockets. Scholars have identified various forms of counter, inverse, or reciprocal surveillance (or sousveillance) (Timan and Albrechtslund 2015). Likewise, three chapters in this book deal directly with the reality that surveillance in public space is often carried out by civilians – through forms of civilian-surveillance that are often intertwined with more liberal public distribution of images and other information captured in public spaces over the internet. This phenomenon encompasses purely "horizontal surveillance" (Newell 2015; see also Marx 2015a, 41; Marx 2015b, 735) between civilians as well as horizontal surveillance that will ultimately be shared with, or accessible to, the state (thus feeding into more vertical, panoptic structures). The examples of civilian-surveillance examined by Hargreaves, Ritsema van Eck, and Trottier, in their respective chapters, all potentially generate heightened horizontal visibility for the surveilled subjects and, particularly in the cases of police reporting apps (Ritsema van Eck) and doxing efforts (Trottier), this visibility can extend to the state as evidence of alleged wrongdoing is shared with (or, at least, becomes accessible and useful to) the police or other state authorities.

Hargreaves examines how people (mostly men) surreptitiously take photographs of others (mostly women) in public places or, alternatively, scour visual street maps (e.g., Google or Bing Maps), and "subsequently upload [the images] to message boards for evaluation and commentary by other users." Trottier, similarly, examines the increasing popularity of "digital vigilantism" – or the practices of making individuals visible online for allegedly committing acts in public space that constitute "legal and moral infractions." According to Trottier, "digital vigilantism involves circulation of information that renders a targeted individual visible on a vast geographic scale, such that the identity of a person engaged in mundane activity in a medium sized Canadian town is accessible beyond that locale with greater scope and speed than could be provided by any other medium."

Finally, Ritsema van Eck describes the emerging phenomenon of "instant reporting" smartphone applications that allow individuals to record or take images of individuals or situations and report them directly, and almost instantly, to the police over cellular or other wireless networks. These "new forms of public surveillance" (Hargreaves), like other mobile forms of surveillance, are particularly subject to the biases of their operators (see, for example, Katell), as the gaze produced "is only possible on the terms of these citizens" (Ritsema van Eck). On the other hand, they also empower the public, as they allow "any citizen on the street" to "record high quality photographs or videos of emergencies and share them with the police instantaneously" (Ritsema van Eck), thus participating in supporting the moral and social imperatives of governments to keep their citizens secure, enforce the law, and provide efficient and effective criminal justice systems. Additionally, as demonstrated by Ritsema van Eck, "exhorting citizens to act as police informants introduces discriminatory biases, pressures towards normalization, and chilling effects." In the case of "creepshots," the activities Hargreaves is concerned about, surveillance and its attendant gaze may also be operated "with a particularly gendered bent." This reality suggests that privacy, on its own, may not be enough to protect the important interests at stake and, Hargreaves argues, our solutions ought to encompass a broader terrain of legal and policy options.

The contributions from Trottier and Hargreaves both provoke questions about the efficacy of legal theories that strip privacy in public spaces – and, in this way, they both add to the critique of such theories offered by proponents of the mosaic theory. Likewise, Véliz argues that the public/private dichotomy (based on spatial considerations, ownership, and property rights) is a poor basis for determining what aspects of our lives should be protected by privacy rights. Indeed, adding many "little brothers" into the mix significantly complicates the set of adequate or ideal regulatory solutions, as "the capture, retention and circulation of identifiable details (such as license plates) through global mediated platforms suggest that the visibility enacted by being located in a public physical space at a specific point in time are augmented by these platforms. As the primary means of social harm employed is mediated, attempts to either maintain anonymity or violate privacy in public spaces are seemingly governed through digital platforms, and amplified by user activity" (Trottier). The fact that someone, somewhere, armed with a camera can bring unprecedented attention to the actions or appearance of any other person, at his or her own whim (whether motivated by mere "interest" or by the aim of making moral miscreants more accountable for their actions), is profoundly altering the possibilities of securing privacy by escaping to the public space and seeking obscurity amongst the many, as suggested by Véliz. In addition, those whose image has been subject to peer-surveillance and digital distribution online may have to accept the perpetual and persistent digital memory of the internet, as "guaranteed permanent removal is a near impossibility – the decentralized nature of the internet actively works against such an outcome" (Hargreaves).

Conclusion

In the end, the distinctions and ideas raised by the authors in this volume may help illuminate possible boundary-marking concepts that can be mobilized to better conceptualize privacy (and the impacts that myriad forms of surveillance have on privacy) and generate new ways of regulating the interactions between privacy and surveillance in public spaces. The interrelated problems raised by the aggregation of numerous data points through long-term or *en masse* surveillance, the involuntary exposure of and emanation of potentially intimate personal information into the public sphere, technological challenges to the prospects of fostering civil inattention and achieving practical obscurity that might compensate for people's increased trackability, and the broadening net of actors participating in surveillance of others in public spaces all point towards the conclusion that we need to find new ways of demarcating the boundaries of privacy, particularly in the streets (and other public spaces), and that we need to do so in ways that are resilient to technological change.

Notes

1 *R. v. Tessling*, [2004] 3 S.C.R. 432, 2004 S.C.C. 67 (Can.).
2 *Id.* at para. 36.
3 See *U.S. Dept. of Justice v. Reporters Committee for Freedom of Press*, 489 U.S. 749, 762 (1989) (practical obscurity); *United States v. Maynard*, 615 F.3d 544, 562 (D.C. Cir. 2010).
4 *United States v. Jones*, 565 U.S. 400 (2012).
5 *Id.* at 420 (Alito, J., concurring), quoting *Kyllo v. United States*, 533 U.S. 27, 34 (2001).
6 *Id.* at 416 (Sotomayor, J., concurring), quoting *Illinois v. Lidster*, 540 U.S. 419, 426 (2004).

References

Arendt, Hannah. 2013. *The Human Condition*. 2nd ed. Chicago, IL: University of Chicago Press.

Balkin, Jack M. 2012. "Room for Maneuver: Julie Cohen's Theory of Freedom in the Information State." *Jerusalem Review of Legal Studies* 6(1): 79–95.

Cohen, Julie E. 2012. *Configuring the Networked Self: Law, Code, and the Play of Everyday Practice*. New Haven, CT: Yale University Press.

Cohen, Julie E. 2013. "What Privacy Is For." *Harvard Law Review* 126: 1904–1933.

Friedman, Batya, Kristina Hook, Brian Gill, Lina Eidmar, Catherine Sallmander Prien, and Rachel Severson. 2008. "Personlig Integritet: A Comparative Study of Perceptions of Privacy in Public Places in Sweden and the United States." *Proceedings of Nordi CHI 2008*, 142–151. Lund: ACM.

Goffman, Erving. 1972. *Relations in Public*. London: Penguin.

Gooas, Kerstin, Michael Friedewald, William Webster, and Charles Leleux. 2015. "The Co-Evolution of Surveillance Technology and Surveillance Practices." In *Surveillance in Europe*, edited by David Wright and Reinhard Kreissl, 51–100. London: Routledge.

Gray, David, and Danielle K. Citron. 2013. "A Shattered Looking Glass: The Pitfalls and Potential of the Mosaic Theory of Fourth Amendment Privacy." *North Carolina Journal of Law & Technology* 14: 381–429.

Habermas, Jürgen. 1989. *The Structural Transformation of the Public Sphere: An Inquiry into a Category of Bourgeois Society*. Cambridge, MA: The MIT Press.

Haggerty, Kevin D., and Richard V. Ericson. 2000. "The Surveillant Assemblage." *British Journal of Sociology* 51(4): 605–622.

Haggerty, Kevin D., and Richard V. Ericson. 2006. "The New Politics of Surveillance and Visibility." In *The New Politics of Surveillance and Visibility*, edited by Kevin D. Haggerty and Richard V. Ericson, 3–25. Toronto: University of Toronto Press.

Hartzog, Woodrow, and Frederic Stutzman. 2013a. "Obscurity by Design." *Washington Law Review* 88: 385–418.

Hartzog, Woodrow, and Frederic Stutzman. 2013b. "The Case for Online Obscurity." *California Law Review* 101(1): 1–49.

Harvard Law Review. 2015. "Considering Police Body Cameras." *Harvard Law Review* 128: 1706–1817.

Himma, Kenneth Einar. 2015. "Why Security Trumps Privacy." In *Privacy, Security and Accountability: Ethics, Law and Policy*, edited by Adam D. Moore. Lanham, MD: Rowman & Littlefield International.

Irion, Kristina. 2015. "Your Digital Home is no Longer Your Castle: How Cloud Computing Transforms the (Legal) Relationship between Individuals and Their Personal Records." *International Journal of Law and Information Technology* 23: 348–371.

Kerr, Orin S. 2012. "The Mosaic Theory of the Fourth Amendment." *Michigan Law Review* 111: 311–354.

Koops, Bert-Jaap. 2014. "On Legal Boundaries, Technologies, and Collapsing Dimensions of Privacy." *Politica e Società* 3: 247–264.

Koops, Bert-Jaap, Bryce Clayton Newell, Tjerk Timan, Ivan Škorvánek, Tomislav Chokrevski, and Maša Galič. 2017. "A Typology of Privacy." *University of Pennsylvania Journal of International Law* 38: 483–575.

Kugler, Matthew B., and Lior Jacob Strahilevitz. 2015. "Actual Expectations of Privacy, Fourth Amendment Doctrine, and the Mosaic Theory." *The Supreme Court Review* 2015: 205–263.

Lyon, David (ed.) 2003. *Surveillance as Social Sorting: Privacy, Risk, and Digital Discrimination*. London, New York: Routledge.

Lyon, David. 2007. *Surveillance Studies: An Overview*. Cambridge: Polity Press.

Marx, Gary T. 2003. "A Tack in the Shoe: Neutralizing and Resisting the New Surveillance." *The Journal of Social Issues* 59(2): 369–390.

Marx, Gary T. 2015a. "Coming to Terms: The Kaleidoscope of Privacy and Surveillance." In *Social Dimensions of Privacy: Interdisciplinary Perspectives*, edited by Beatte Roessler and Dorota Mokrosinska, 32–49. Cambridge: Cambridge University Press.

Marx, Gary T. 2015b. "Surveillance Studies." In *International Encyclopedia of the Social & Behavioral Sciences* (2nd ed.), edited by James Wright, 23: 733–741. Oxford: Elsevier.

Miller, Kevin. 2014. "Total Surveillance, Big Data, and Predictive Crime Technology: Privacy's Perfect Storm." *Journal of Technology Law & Policy* 19: 105–146.

Moore, Adam D. 2010. *Privacy Rights: Moral and Legal Foundations*. University Park: Pennsylvania State University Press.

Newell, Bryce Clayton. 2015. "Transparent Lives and the Surveillance State: Policing, New Visibility, and Information Policy" (doctoral thesis, University of Washington). https://digital.lib.washington.edu/researchworks/bitstream/handle/1773/33984/ Newell_washington_0250E_14460.pdf.

Ready, Justin T., and Jacob T.N. Young. 2014. "Three Myths about Police Body Cams." *Slate*, September 2. www.slate.com/articles/technology/future_tense/2014/09/ferguson_body_cams_myths_about_police_body_worn_recorders.html.

Roberts, Andrew. 2015. "A Republican Account of the Value of Privacy." *European Journal of Political Theory* 14(3): 320–344.

Selinger, Evan, and Woodrow Hartzog. 2014. "Obscurity and Privacy." In *Routledge Companion to Philosophy of Technology*, edited by Joseph Pitt and Ashley Shew. Pre-press version available at http://ssrn.com/abstract=2439866.

Sennett, Richard. 1992. *The Fall of Public Man*. New York: W.W. Norton & Company, Inc.

Slobogin, Christopher. 2012. "Making the Most of United States v. Jones in a Surveillance Society: A Statutory Implementation of Mosaic Theory." *Duke Journal of Constitutional Law & Public Policy* 8: 1–37.

Solove, Daniel, Marc Rotenberg, and Paul M. Schwartz. 2006. *Information Privacy Law*. New York: Aspen Publishers.

Timan, Tjerk, and Anders Albrechtslund. 2015. "Surveillance, Self and Smartphones: Tracking Practices in the Nightlife." *Science and Engineering Ethics* (online first, advance publication). DOI: doi:10.1007/s11948–11015–9691–9698.

Tonkiss, Fran. 2003. "The Ethics of Indifference: Community and Solitude in the City." *International Journal of Cultural Studies* 6(3): 297–311.

Warren, Samuel D., and Louis D. Brandeis. 1890. "The Right to Privacy." *Harvard Law Review* 4: 193–220.

Chapter 2

In the privacy of our streets

Carissa Véliz

Introduction

If one lives in a city and wants to be by oneself or have a private conversation with someone else, there are two ways to set about it: either one finds a place of solitude, such as one's bedroom, or one finds a place crowded enough, public enough, that attention to each person dilutes so much so as to resemble a deserted refuge. Often, one can get more privacy in public places than in the most private of spaces. The home is not always the ideal place to find privacy. Neighbours snoop, children ask questions, and family members judge. When the home suffocates privacy, the only escape is to go out, to the coffee shop, the public square. For centuries, city streets have been the true refuges of the solitaries, the overwhelmed, and the underprivileged.

Yet time and again we hear people arguing that we do not have any claim to privacy while on the streets because they are part of the so-called public sphere. The main objective of this chapter is to argue that privacy belongs as much in the streets as it does in the home.

In the first section, I argue that the distinction between the private and the public sphere does not map onto what should be guarded by privacy. In other words, we cannot figure out what ought to be protected by privacy by referring to the private/public distinction. In the second section, I argue that privacy can only be safeguarded by protecting both our personal information and our autotopos – a metaphorical sensorial personal space or privacy bubble – and offer a few reflections on the value of privacy. The third section is about the streets of a city as a vital commons, the importance of privacy to ensure their full enjoyment by all citizens, and the threats that are eroding privacy in public spaces. The fourth section concludes the chapter with a final defence for privacy in the streets.

The street and the public sphere

Because of the close etymology between the noun *privacy* and the adjective *private*, it is understandable to assume that privacy will always be on the "private" side in the dichotomy between the private and the public. Whether this is true depends on how we conceptualize the distinction. When it comes to streets and public squares, we seem to think these belong to the public sphere on account of two common ways of dividing the public from the private: on the basis of physical spaces and on the basis of funding. As I shall argue, on these ways of separating the public from the private it is untrue that what ought to enjoy privacy always belongs to the private sphere.

For the time being I will not offer a philosophically rigorous definition of privacy. As a rough and ready working understanding of privacy, it can be said that the private, in the sense of privacy, refers to that which is typically kept hidden by an individual from most other people in a certain society. I will mostly rely on paradigmatic privacy cases to argue that, whatever definition of privacy we favour, and as long as we want it to roughly accord with common usage, we cannot rely on the private/public distinction as it is commonly understood in order to justify a claim about what ought to be kept private.

The private as a physical space

One commonsensical way to divide the private and the public is to suppose that the private sphere is a spatial zone intimately related to individuals and families that is off-limits, or of no concern, to the government or to other citizens. That it is of no concern to others is meant normatively, not descriptively. A neighbour might be very interested in what you look like without your clothes on behind the shut curtains of your bedroom, but he should not be (or, at the very least, he should not have access to your bedroom). The paradigmatic locus of the private in this sense is the home.

By contrast, the public sphere is a shared zone that is the legitimate concern of both the government and the people, the citizenry in general. The public sphere is an area that is subject to the authority of the government. Paradigmatic public places are streets, public squares, parks, and government buildings.

One might think, then, that under this division, people can expect privacy in the guarded comfort of their houses (as long as they do not engage in criminal activities). In the streets, however, privacy is not to be expected or respected. Thus, there are those who think that even if photographers do not have the right to take shots of people who are in their homes, as soon as people step out of their houses, it is fair game to capture them with a lens. Photographer Nick Turpin, for example, believes that "what happens in a public place should be a matter of public record" (Laurent 2013). Turpin acknowledges taking photographs may have bad consequences. "I could be photographing a couple kissing while they shouldn't be kissing. But if they are doing it in a public space, it's a risk that they're running ...", he said.

Nick Turpin's example seems to provide strong support for his position because kissing on the street appears to be a particularly voluntary act, but not all behaviour seems so deliberate. Consider the case of an individual getting hit by a car on the street. When the paramedics arrive, they cut through his clothes in order to give him medical treatment. In this case it is much less clear that witnesses are entitled to take photographs of him naked and injured – much less publish them. The fact that he did not intend to be unclothed in public and the sensitive nature of the situation are reasons for people to act with discretion. Cases such as these are enough to show that Turpin is wrong, that people can have a claim that others respect their privacy when being in a public space – at least in certain circumstances.

Even in cases in which it seems that people have deliberately chosen to carry out a sensitive act such as kissing in public, it is worth noting that often our "choices" are much more constrained than they seem. A couple may not have the financial means to live in places where they can enjoy privacy or to pay for a hotel room. A political dissenter may not be able to meet with a fellow activist in her home because she suspects her house is under surveillance. Sometimes, the only breaths of privacy we can get are precisely in public spaces, away from people who can recognise us.

In this vein, Patricia Meyer Spacks notes how women often find privacy only outside of the home: "the housewife wants privacy specifically to get away from her family for a time" (2003, 1). The home is thought to provide a temporary relief from the demands of the larger society. At home, one can stay in one's nightgown, put one's feet on the table, and relax. Sometimes, however, domestic chores and family obligations can get in the way of one's individual privacy. Having to engage with family can be an obstacle to having the chance to spend time by oneself writing in one's diary, for example, or speaking privately with a friend, away from the eyes and ears of one's spouse and children. With the passage of time, intimacy can become a burden, and caring can become oppressive (Moore 1984, 277, 42). If the house is too small, noises from others' activities can be annoying. An escape from others is important for wellbeing. Men, women, children, and teenagers all get privacy from their families by leaving their house for some hours every day and entering the public sphere. School, work, and solitary walks (even if they take place in crowded streets) can all serve as potential privacy resources.

These cases show that sometimes privacy sides with the public in the private/public divide. Above and beyond these counterexamples, there is an equally important limit to be noted with respect to the public and the private as normative spatial zones. One problem with separating the private and the public on the basis of that which is the legitimate concern of individuals and families or the government and the citizenry is that it is a normatively laden demarcation. Such a characterisation could not explain why something should be protected by privacy; it would be question-begging. In other words, it is not enough to say "*x* should be protected by privacy because *x* is part of the private domain." One would have to then justify why *x* is off-limits or not a legitimate concern of the government or the public.

The private as that which is not publicly funded

Another common way of drawing a line between the private and the public is by appealing to the sources of funding. If an institution or a space is funded with money earned by businesses or individuals acting in a non-governmental capacity, it is usually thought to belong to the private sector. According to this way of dividing the pie, individual citizens fall into the private domain on account of their wages, estates, and shares being personal assets. If an institution or a space is funded with money coming from taxpayers, it is usually listed as belonging to the public sector. On this view, streets would be considered public because they are built with government funds.

One may think that privacy has nothing to do with what is publicly or privately funded, but it is plausible to think that paying for something (even if only partially) earns you certain kinds of entitlements. Thus, one might think that just as employers are entitled to some degree of knowledge about their employees (at least while the latter are on duty), so citizens are entitled to keep an eye on their public officials because they are paying for their salaries by way of their taxes. In the same vein, it is plausible to think that people are entitled to use the streets as they wish – e.g. taking photographs of others or engaging in other privacy invading practices – because the streets are theirs, since they pay for them through their taxes.[1]

The first thing to point out is that financially separating the private and the public is problematic. When it comes to spaces, there are some public squares that are privately owned (e.g. Zuccotti Park, in New York City, where the Occupy Wall Street camp was placed), and some spaces that are publicly funded but used privately (e.g. public housing).

Even if we could sharply distinguish between the private from the public on account of funding, marking the divide by focusing on the origin of money will not help us in deciding what ought to be kept private. It is clear that whether spaces are publicly or privately funded is no good guide for whether people should enjoy privacy when using them. Patients are entitled to high degrees of privacy regardless of whether they are treated in a public or a private hospital, for example. Likewise, people should be able to enjoy privacy in their bedrooms even if they are staying for free at a house owned by the government.

The implication of these counterexamples is that what was true of the spatial delimitation between the private and the public is also true of a financial delimitation: privacy cannot be descriptively specified or normatively decided by reference to what is private in a financial sense. The implication for privacy in the streets is that the fact that streets belong to the public sphere on account of them being built with government funds cannot justify not being able to enjoy privacy while making use of them.

If appealing to the public and private distinction is unhelpful, how can we know what privacy is and what degree of privacy should we expect while on the street? In what follows I offer an account of privacy that helps determine when and where privacy ought to be respected.

Privacy and its value

Let us take a situation in which Tom peeps through Dana's bedroom window as a paradigmatic example of a loss of privacy. Precisely because privacy in public spaces is a contested issue, it is better to take as a paradigm case one for which most people would have no objections in seeing it as a privacy invasion in order to come up with a definition of privacy which can then by applied to public spaces. The peeping Tom paradigm encapsulates many lessons about privacy.

One of those lessons is that privacy is not only about information. Dana might worry that Tom may gather sensitive information from watching her bedroom (e.g. who she sleeps with). But suppose Dana is simply reading the newspaper in her bedroom and she is not worried about the (trivial) information Tom can glean by spying on her through the window. Dana might still be upset about Tom peeping through her window, even when she is not concerned about the information Tom is learning.

The unease experienced when someone looks at you when you do not want to be looked at does not seem to be captured by describing it in terms of the information others may acquire about you. Indeed, if Dana were to realise that Tom is looking through her window (and supposing there is nothing she can do to make him go away), she might be so disturbed that she might find she can no longer concentrate on her reading. What matters, then, is not only information, but also sensorial access. We care about others being able to see us, hear us, touch us, smell us.

Another valuable lesson is that culture and social norms matter. In our societies, the bedroom is a culturally designated area of privacy – an area where it is generally recognised that individuals have a particularly strong claim to privacy.

The home and the bedroom are paradigm cases of private spaces, but I contend that social norms are such that we implicitly do and should acknowledge a kind of bubble of personal space that surrounds us. The bubble expands or shrinks depending on the social conventions accorded to different places and situations. I call this bubble the *autotopos.*[2] Autotopos, then, refers to a metaphorical sensorial personal space, a sensorial self-space. It is the kind of sensorial space that people in a society commonly would not want anyone, other than him/herself (and perhaps a very limited number of other people chosen by him/her), to access.

There are two ways our autotopos can be accessed. First, when someone sensorially enters a culturally established personal zone of ours. That is, when someone (through direct or indirect perception such as cameras and microphones) sees, hears, smells, or touches us in a zone where there are cultural expectations to be free from the eyes, ears, touch, and presence of others (e.g.

in the toilet). Second, our autotopos is accessed when we are witnessed engaging in some activity or being the subject of some event that typically evokes the desire to have no witnesses or very few chosen witnesses (e.g. being naked). Such an activity or event may happen in the public sphere, as in the case in which paramedics strip a person who has suffered an accident in order to offer them medical attention while on the street. The fact that someone is in a public space should not give others the impression that they can invade her privacy; particularly if she is in a sensitive situation.

The autotopos is metaphorical because it does not precisely equate with physical or territorial space, even if the two might overlap. It is a *sensorial* bubble. Placing a camera outside of your bedroom window may not invade a physical space, but it does invade a sensorial space. Placing a foot on a land that you have never visited and yet own halfway across the world may be an invasion of private property, but not of your privacy. For privacy, what matters is that one can be seen, heard, smelled, and touched by others.

It can be said, then, that Dana possesses privacy with respect to Tom if both her autotopos and her personal information remain unaccessed by Tom.

By personal information I mean the kind of information that is common for people in a certain society not to want anyone, other than him/herself (and perhaps a very limited number of other people chosen by him/her), to know about. Personal information also includes information that someone is particularly sensitive about and has taken measures to conceal from others.

According to this account, Tom accesses Dana's personal information the moment Tom gains knowledge of it. In other words, it is not enough for the information to be accessible to Tom. If Tom finds Dana's diary but does not read it – as long as the diary remains unaccessed – Dana has lost no privacy.[3]

One might wonder what makes the two species – informational and sensorial access – part of the genus of privacy. The unity of the category of privacy is founded on the notion of being personally unaccessed and the kinds of interests we have in not being accessed by others. Privacy protects us from a) certain kinds of harms that may come about as a result of other people having access to our personal life, b) the demands of sociality, c) being judged and possibly ridiculed by others (and thus from self-conscious negative emotions such as shame and embarrassment), and d) the discomfort of being watched, heard, and so on. Privacy is valuable because we value the interests it protects.

Through such protection, privacy enables individuals to present themselves in different ways to different people (Marmor 2015). In his book, *The Presentation of the Self in Everyday Life* (1959), sociologist Erving Goffman described how this control was crucial for successfully managing both professional and personal relationships. Similarly, James Rachels (1975) has pointed out that being able to control who has access to oneself and information about one is intimately related to our ability to maintain different kinds of relationships.

It could be objected that we should not be given the chance to act differently in the presence of different people because that amounts to being two-faced, dishonest, or inauthentic. On this view, a lack of privacy would implement absolute transparency and that would be good, because it would force people to show their true colours to everyone alike, or so the thought goes. As Rachels has argued, however, different relationships are partly defined by different patterns of behaviour. One's best friend would not be one's best friend if one could not cry, swear, and express one's fears in her presence. If one were to display the same kind of behaviour with, say, one's students, one would be acting inappropriately. It is not that one is being dishonest when one acts with one's student differently than when one is the company of friends. Rather, relationships function as a kind of division of labour, and loading one kind of relationship with the burdens of another type of relationship creates confusion and dissatisfaction.

Furthermore, as Thomas Nagel (1998) has pointed out, these patterns of behaviour are not dishonest because they are not meant to deceive; they are social conventions that are typically shared and well known by all. Being able to conceal certain aspects of ourselves in order to present ourselves in appropriate ways depending on the public we are facing "protects one from the sense of exposure without having to be in any way dishonest or deceptive, just as clothing does not conceal the fact that one is naked underneath" (Nagel 1998, 8).

The more privacy we can have in the streets, the more we will be spared from unwanted gossip, and the more possibilities we will enjoy of presenting ourselves in different ways. Privacy in the streets is important for people who cannot be themselves elsewhere. It is valuable because it allows people to explore things, places, and people that they otherwise would not explore because of outside influences, for fear of what others might think of them, or for fear of repercussions.

In the streets, privacy keeps us safe from various harms. The more personal information others know about us, the more likely it is that it will be used against us in unfair ways. Criminals can use information about us, such as where we live, to commit identity theft, for instance, and businesses can use that information to create personalised pricing, making some people pay more than others depending on where they live.

Enjoying privacy in the streets also enables us to have time and space for our thoughts. Walking in the midst of a crowded street in which no one pays attention to us and where we do not have to spend energy attending to others can be a respite from the burdens of social interaction (Gavison 1980).

In the political realm, privacy contributes to the protection of liberal, democratic, and pluralistic societies (Gavison 1980). Privacy fosters and encourages individuals' autonomy by shielding them from external interference. Political liberty requires that people have the right to keep private their votes, associations, and thoughts, if they so wish. When democracies corrode and disagreeing with power becomes dangerous, privacy protects political dissenters such as activists and whistleblowers. When advocating against injustice may endanger one's life or one's loved ones, having the possibility of anonymous protest and resistance becomes crucial to defend democratic ideals.

With the right social norms and laws in place, streets can be rich privacy resources where individuals can take refuge to explore the world around them, to be alone with their thoughts or in the company they choose, to be anonymous, to be themselves, to be safe surrounded by strangers, to protest for just causes. If we let privacy die in public places, on the other hand, the streets can become inhospitable places where passers-by, through venturing out of their homes, risk stress, crime, loss of their reputation, and more.

Rural public places are outside the scope of this chapter. I focus on city streets because already, most of the world's population is urban, and this trend is on the increase, with some experts estimating that by 2210, nearly 87% of people will be living in cities (Fuller and Romer 2014).[4] Furthermore, protecting privacy is much more of an issue in cities, and increasingly so. In rural areas, privacy challenges are much the same as they have been for hundreds of years, with some exceptions provided by technologies such as smartphone cameras and drones (which also affect cities). The scope for action in modifying privacy risks in rural areas is more limited, constrained by defining features of geography and density of population. When it comes to cities, in contrast, it is widely up to society how to fashion streets and technology to accord with our privacy desires. The city is utterly suffused with technological artefacts that put privacy at risk: from CCTVs to smartphone cameras, drones, electronic cards, smart meters, and more.

Numerous cities around the world are becoming "smart", one step at a time. Santander, Amsterdam, Barcelona, and Dublin have all launched projects to further digitize urban experience. The most radical and ambitious project is currently being undertaken by Google's Sidewalk Labs in Toronto. Their objective is to transform 800 acres of waterfront property into the most technological urban area the world has thus far seen (Badger 2017). With smarter cities, innumerable privacy challenges are coming our way. The better we understand the value of privacy in our streets, the better prepared we will be to face those challenges. In the following section, I further explore the value of being able to fully enjoy the streets and some of the threats to privacy in public places.

Privacy in the streets

Imagine being barred from stepping onto the streets. Consider a city life constrained to spending it indoors, and in privately owned open spaces such as gardens. Other things being equal, it seems that this would be an impoverished life. It is better to be able to go out into the street than not to have that possibility. In other words, we gain in cultural richness, valuable opportunities, and pleasurable experiences by being able to enjoy the streets of our cities. Streets provide us with the possibility of breathing fresh air, meeting both friends and strangers who may become friends, enjoying outdoor cultural events such as street music and theatre, participating in the cultural milieu of our city by appreciating trends related to fashion, architecture, etc. They enable easy access to shops and they allow pedestrians to discover and enjoy new places and people.

Furthermore, except for people who are wealthy enough to pay others to do some tasks, going out into the street is usually not optional. It is not only that streets are full of wondrous possibilities; using them is necessary for the livelihoods of most people. One can hardly avoid streets if one needs to go to work at an office, get groceries (online services and delivery are still limited; for instance, they are less immediate than buying something at a local shop), or enjoy some much needed human contact.

Because of their ubiquity and their ability to connect parts of the city, streets are one of the most fundamental commons experienced in an urban context. Streets are the veins of a city, and citizens are the oxygen that runs through them, making the metropolis live and breathe. However, there are certain preconditions necessary for citizens to be able to fully enjoy their streets. Streets must be reasonably safe. If people are afraid for their physical integrity when walking down avenues and roads, they will likely restrict their movements, the places they visit, and the times they go out at, thus not being able to fully enjoy the commons. Likewise, if streets are not clean enough, or if they are not accessible enough (for the handicapped, the elderly, for parents with pushchairs), people are more likely to avoid them. In a similar vein, if people cannot enjoy some degree of privacy on the streets, if they are constantly exposed to invasions, they are likely to go out less than they otherwise would, or to alter their habits in protective ways that are detrimental to their enjoyment of the streets. For people to feel safe, both their personal information and their autotopos have to enjoy a reasonable degree of respect.

The documentary *$ellebrity* (Mazur 2012) illustrates how traumatic a lack of privacy can be in public spaces. It tells the story of celebrities reduced to tears and anxiety because they cannot take a step outside their homes without having paparazzi harass them with questions and pictures. The lack of privacy can be so unnerving that it can drive people to take unnecessary risks: although it is a matter of controversy, for years it has been suggested that Princess Diana's death in a car crash may have been caused by her chauffer trying to escape paparazzi. The issue of what degree of privacy should celebrities enjoy in public spaces is complex and beyond the scope of this chapter. Most people will presumably agree, however, that whether it is the right price to pay or not for fame, the inability to be anonymous in the street is a serious loss, and ordinary people, however we may want to define them, should not have to suffer such loss. Of course there is a certain vulnerability inherent to stepping onto the sidewalk. It is unreasonable to expect complete privacy on the streets. Leaving the enclosed space of the home necessarily exposes us to a myriad of people, with all the risks that involves. By going out of our homes we run both physical risks, such as getting hurt by cars or other pedestrians, and privacy risks (e.g. others may notice us when we would prefer not to be noticed). Our autotopos is bound to be at risk by others looking at us, and so is our personal information, as at least some passers-by or neighbours are bound to notice where we live as we go out of our homes.

An important distinction should thus be made between inevitable kinds of exposure and unnecessary invasions of privacy. Inevitable exposure comes from the need to make the streets available to the enjoyment of all. Other things being equal, people cannot be banned from walking down any particular street without seriously impeding their enjoyment of public spaces, and likewise it cannot be forbidden to look at other people who are on the street. There is thus an unavoidable risk that one may bump into people one would wish not to see or not to be seen by. As Goffman points out, public spaces render "persons uniquely accessible, available, and subject to one another. Public order, in its face-to-face aspects, has to do with the normative regulation of this accessibility" (1963, 22). In cases of unavoidable exposure, social norms can go a long way towards minimising negative consequences. We tend to practise decorum in cases in which we perceive we might be invading someone's privacy.

People "divulge information" about themselves, often against their wishes, by their "mere presence in a situation" (Goffman 1963, 103). Imagine seeing a work acquaintance walking down the street. He is holding someone's hand. As soon as he notices you, he blushes, lets go of his companion's hand, and faces a shop window, pretending not to have seen you, giving his back to you as you pass by. People wanting to control others' access to them typically avert their eyes to avoid eye contact (Goffman 1963, 93). The tactful approach to this situation – given that your acquaintance clearly does not want to interact with you – is not to force an encounter, join him in pretending you did not see each other, walk by, and not talk about the incident with others.

As has been mentioned, context is partly what modulates the boundaries of people's autotopos. Helen Nissenbaum's framework of contextual integrity helps explain the importance of different situations for privacy. According to her, what is most important to people is not simply limiting access to personal information, but rather ensuring that information flows *appropriately*. Context is what determines that appropriateness. (Nissenbaum 2010, 1–4). As Nissenbaum explains, "[c]ontexts are structured social settings characterized by canonical activities, roles, relationships, power structures, norms (or rules), and internal values (goals, ends, purposes)" (2010, 132). Social norms regulating behaviour on the streets are thus influenced by culture and by nuances in the situation (e.g. whether the person we encounter is accompanied, whether she may feel embarrassed about the way she is dressed or the store she is visiting, etc.).

It may not always be easy to know how much privacy someone wants if one happens upon her on the street. Social cues such as avoiding eye contact help in communicating privacy needs; they are clues as to whether someone needs space from us, whether their autotopos needs to be respected at a certain time and place. As Irwin Altman puts it, "[p]eople implement desired levels of privacy by behavioural mechanisms such as verbal and paraverbal behavior, nonverbal use of the body, environmental behaviors and cultural norms and customs" (1975, 17).

Social norms that minimise inevitable exposure include not watching people intently, and not following them around. In other words, social decorum in public places demands of people that they interact in an "unfocused" way. For example, that they manage copresence on the streets by quickly glancing at others, rather than staring (Goffman 1963, 24). There is a kind of courtesy involved in giving a passer-by "enough visual notice" to acknowledge his presence "while at the next moment withdrawing one's attention from him so as to express that he does not constitute a target of special curiosity." Goffman calls it "civil inattention," a kind of "dimming of lights" when persons pass each other on the streets (1963, 84).

In short: in the streets we are accessible, but what matters the most is that we are not accessed in ways that make us uncomfortable. Because we are accessible, a healthy and full use of the streets depends on trusting that others will not access us inappropriately, even when they could.

Avoidable invasions of privacy violate the respect for others' personal information and autotopos. They are practices that are not necessary for all to enjoy the streets and that diminish the privacy of individuals. Avoidable invasions of privacy may come from other individuals, businesses, and governments.

Other passers-by can invade our privacy by taking photographs of us and publishing them. When photographs are taken surreptitiously, there is always a risk of the target's autotopos being invaded, as the photographer usually does not have enough information to know whether the target might be in a certain context where she does not wish to be photographed (e.g. perhaps she does not want people to know she was where she was). The invasion can be especially damaging if the photograph portrays what is clearly a sensitive situation, such as the victim of an accident or some other kind of disaster, or a couple kissing. Photographs can be especially harmful because they can involve both a breach of people's autotopos (which can lead to people feeling embarrassed from their image being exposed, above and beyond issues of information) and they can also disclose very personal information (e.g. a romantic relationship). The popularisation of smartphones has made photographing in the streets a much more common practice. Social media sites like Facebook have made it easier to publish pictures to a wide audience without always knowing whether it might have bad consequences for someone. Wearable cameras such as Google Glass or MeCam represent further threats to privacy.

Businesses can invade our privacy by using facial recognition systems and by identifying people through their smartphones and following their every step when they enter a shop (Clifford and Hardy 2013). Shops are using the Wi-Fi signals from mobiles to fish for personal information about shoppers and to track their buying behaviours,[5] and they use cameras and facial recognition to assess people's moods, among other things. Commercial invasions of privacy are mostly about personal information, even when they use people's autotopos (as in the case of facial recognition) to get to the former. This information is then used for marketing purposes, but it also gets sold to data brokers (Hoofnagle 2016). Data brokers then sell profiles to the highest bidder; they have been known to sell lists of rape victims, AIDS patients, and more (Hicken 2013).

Finally, governments can invade our privacy through the use of technology. The most obvious example are CCTV cameras. It can be argued that surveillance cameras should be considered as an inevitable kind of invasion because they ensure safety and are thus necessary for streets to be enjoyed by everyone. However, evidence regarding the effectiveness of cameras at deterring crime is contradictory, with a significant amount of it suggesting no effect whatsoever (Cameron et al. 2008). Other evidence suggests that they are only as effective as streetlights (Welsh and Farrington 2004), yet much more privacy-invading. In any case, even if we agree on their necessity, there are ways in which they can be abused: by keeping footage for longer than necessary, by tracking individuals without reasonable suspicion, by selling that footage, or by irresponsibly publishing images. In 1995, for example, a CCTV camera caught a man about to slash his wrists. The footage was then shown to more than nine million viewers on a BBC show without concealing his face (Dyer 2003).

A less well-known technology through which governments invade citizens' privacy are International Mobile Subscriber Identity-catchers, or IMSI-catchers – fake cell phone towers that trick mobile phones into connecting to them. Once connected, IMSI-catchers can collect identification and location data, as well as enable eavesdropping on phone conversations, text messages, and web browsing (Hicken 2013). There is evidence that this equipment is being used by the police across London to spy on people, for example, at peaceful protests and near the UK parliament (Bryant 2016).

Popular advice given online to protect one's privacy now includes leaving one's phone at home when going to a protest (Dunne 2017). Phones guard some of the most personal information people hold about themselves, and protestors are not keen on the police having information on them that could potentially be used against them. However, not having a phone during a protest is a significant difficulty. Having a phone is important for protestors to feel safe. Protests can get out of control, and friends and loved ones often get separated, sometimes simply as a result of the number of people protesting, and sometimes as a result of police intervention or violent disruptions of various kinds. Having a phone allows people to be in touch, reorganise, and find each other. Having a phone also means that one is able to call for help in case of injury or arrest. Sometimes activists are forced to stay more time in police custody than necessary because they could not alert their loved ones or their lawyer that they had been arrested. Finally, having smartphones allows for the possibility of filming police wrongdoing. People having to leave smartphones at home, or worse, not going to protests for fear of invasive practices by the police, are not good signs of healthy democracies.

With all of these avoidable invasions of privacy, social norms such as decorum are not enough to guarantee privacy – appropriate laws must be passed. It may be argued that laws are not necessary because most of these privacy invasions do not threaten the use of the streets, as people do not seem to care about them, or not enough to stop them from stepping onto the sidewalk. There are at least three responses to this objection.

First, people may be unaware of many of the privacy invasions they are being victims to. By deceiving victims of privacy invasions, privacy offenders are failing to respect people's autonomy. Covert surveillance deceives victims about their world – it controls the victim's beliefs about whether she is being watched (Benn 1971). These false beliefs affect the victim's desires and her actions. She acts differently than she would if she knew she was being watched. Through deception, watchers thwart citizens' attempts to make rational choices for themselves. For people to be able to autonomously decide how to lead their lives, they must be reasonably well-informed about whether someone is watching them, among other things (Benn 1971).

Second, people may not have a choice but to go out onto the street in order to lead their lives. Their continued use of the streets may not be taken as consent or agreement. People may feel so impotent in the face of so many privacy threats that they may resign themselves to the status quo. It is impossible to protect oneself from all the privacy invasions that take place regularly on our streets, and resigning oneself to this fact is not the same as not minding or not having a strong preference for privacy. Furthermore, the discomfort of not enjoying privacy might still be influencing people's behaviour for the worse, even if it does not go so far as to make people not go out at all. It is hard to tell whether people might be expressing less affection in public, for example, because they feel more watched than in the past, or whether they might be avoiding certain places that they would not avoid were they to enjoy more privacy.

Third, even if people are informed of the privacy invasions they are subject to and they still do not mind them, there is still reason to think that there is a moral wrong happening, that the street as a commons is being damaged. It is possible that privacy damages are similar to ecological damages in that they are often noticed when it is too late. Individual and small acts of littering or polluting do not seem to make a difference. Yet acts of littering and polluting accumulate, and there comes a moment when the damage is noticeable, great, and often irreversible. Similarly, every bit of personal data that we give away or that gets invaded seems irrelevant: it does not hurt and people may not notice it. But as we lose more and more privacy, as more information about us becomes unprotected, we become more and more vulnerable to harms, and the likelihood and number of privacy harms is bound to go up. We can already appreciate this trend. According to the Federal Trade Commission (2016), for example, identity theft complaints in the United States increased by 47% between 2014 and 2015.

There is a further question about what kinds privacy invasions should generate legal liability.[6] Minor privacy invasions such as stares should clearly not be legally regulated, unless we want to build a police state that would, ironically, only erode privacy further. The issue merits a chapter on its own and is beyond the scope of this one, but some observations are in order.

Although respecting other people's privacy includes respecting information they are sensitive about, it is not appropriate for the law to legislate according to individual sensitivities. If you know someone is sensitive about something seemingly irrelevant such as their height, the virtuous thing to do is not to try to find out how tall they are. However, the law should not punish such privacy invasions, as people can be sensitive about too many things, some of them unreasonably mundane, and it is hard to ascertain whether they have taken steps to conceal that information, and to what extent gaining access to that information has been a privacy invasion. Personal sensitivity should be taken into account, morally, but is much too subjective to be legally enforced. Ordinary people would suffer too many restrictions from this kind of legislation. If, however, someone retrieves information that a victim is sensitive about through a clear privacy invasion, such as breaking and entering into her bedroom, then that breach should be punished – not in virtue of the kind of information retrieved, but in virtue of the kind of invasion used to retrieve the information (breaking without consent into a legally established zone of privacy).

What can be legally enforced are well entrenched and generally recognised norms regarding privacy. The theft of personal information from mobile phones, for instance, should be forbidden – particularly when it comes to businesses. Businesses should not be able to profit from violating customers' personal information without their consent, and personal information should never be used against the interests of customers (e.g. by driving prices up depending on where a person lives, or what she does for a living).

Legal regulation should be heavier on the side of businesses than individuals, and it should always be in the service of citizens' wellbeing. Being able to enjoy the streets without fear of undue exposure is an important component of the wellbeing of populations. Through social norms and laws, access must be as limited as possible (without encroaching on others' claim to enjoy public spaces) to both personal information and people's autotopos while they enjoy the streets. As things stand, there is much room for improvement.

Conclusion

In this chapter I have argued that the public nature of streets does not justify a radical lack of privacy. Both political philosophy and morality are in the business of figuring out what a better world would look like. I have argued that a world in which we can enjoy more privacy in the streets would be a better one.

We may go on eroding privacy in the streets and may not notice all of its effects immediately. At the moment, only some people are bearing the brunt: the unlucky ones who suffer identity theft, the ones who live in dangerous countries and risk getting kidnapped, LGBT people living in intolerant societies, people engaging in forbidden love, and so on. The risk is that, when those effects reach most of us, when we are all made to suffer grave consequences from lack of privacy, privacy invading customs and practices may be so entrenched in our culture, economy, and laws, that it may be very difficult to go back to a more private world.

Once privacy in the streets is lost, only the very rich will be able to secure it. Only people who can afford to pay for other people to go out into the streets to take care of everyday chores for them, and for properties that can offer them privacy, will be able to fully enjoy the many benefits of privacy. *The Guardian* recently reported that, for the first time, this generation will be poorer than its parents (Barr and Malik 2016). In a historical moment in which housing prices seem to be soaring in much of the world, in which a significant proportion of young professionals are either forced to live with their parents or to share a flat with others, being able to enjoy privacy in the streets becomes all the more important. Better to decide what kind of streets we want now, while we are still in time to shape social norms and laws to be more respectful of privacy, than to wait until we feel the full force of a lack of privacy – the damages may be severe, and it may be too late to turn back.

Acknowledgement

Some parts of the first two sections have been taken from the author's dissertation (Véliz 2017).

Notes

1 Note that this kind of argument could be reversed as a point against government surveillance. Since it is the people who pay and own their streets, they should be able to enjoy the streets as they wish – and they might wish to make use of them without the discomfort of being watched by surveillance cameras. This argument, however, is just as unsuccessful for, as we will continue to see, the source of funding cannot justify whether something ought to be protected by privacy.

2 My thanks to Roger Crisp for suggesting this term.

3 The adjective "unaccessed" is not found in any dictionary, but there is no suitable existing term to convey in one word the property of not having been accessed. "Inaccessible" denotes the property of not being able to be accessed, which is different from being accessible yet not actually accessed. Analogous differentiations exist in English, however, that use the same prefixes (e.g. indisputable/undisputed, inalterable/unaltered, etc.).

4 Calculation made by the author based on data provided in Fuller and Romer 2014, 6.

5 To minimise others' ability to track one, it is advisable to turn off the Wi-Fi on one's smartphone when leaving one's home.
6 My thanks to an anonymous reviewer for pointing out this issue.

References

Altman, Irwin. 1975. *The Environment and Social Behavior. Privacy, Personal Space, Territory, Crowding*. Monterey: Brooks/Cole Publishing Company.

Badger, Emily. 2017. "Google's Founders Wanted to Shape a City. Toronto Is Their Chance." *The New York Times*, October 18. www.nytimes.com/2017/10/18/upshot/taxibots-sensors-and-self-driving-shuttles-a-glimpse-at-an-internet-city-in-toronto.html.

Barr, Caelainn, and Shiv Malik. 2016. "Revealed: The 30-year Economic Betrayal Dragging Down Generation Y's Income." *The Guardian*, March 7. www.theguardian.com/world/2016/mar/07/revealed-30-year-economic-betrayal-dragging-down-generation-y-income.

Benn, Stanley. 1971. "Privacy, Freedom, and Respect for Persons." In *Nomos XIII: Privacy*, edited by J. Roland Pennock and John W. Chapman. New York: Atherton Press.

Bryant, Ben. 2016. "VICE News Investigation Finds Signs of Secret Phone Suveillance Across London." *VICE*, January 14. https://news.vice.com/article/vice-news-investigation-finds-signs-of-secret-phone-surveillance-across-london.

Cameron, Aundreia, Elke Kolodinski, Heather May, and Nicholas Williams. 2008. *Measuring the Effects of Video Surveillance on Crime*. Report to the California Research Bureau. Available at www.popcenter.org/library/scp/pdf/210-Cameron.pdf.

Clifford, Stephanie, and Quentin Hardy. 2013. "Attention, Shoppers: Store Is Tracking Your Cell." *The New York Times*, July 14. www.nytimes.com/2013/07/15/business/attention-shopper-stores-are-tracking-your-cell.html?pagewanted=all&_r=0.

Dunne, Carey. 2017. "Ten Easy Encryption Tips for Warding Off Hackers, the US Government – and Russia." *Quartz*, January 4. https://qz.com/873922/easy-encryption-methods-how-to-improve-your-online-security-with-encryption-tips-from-a-cryptoparty/.

Dyer, Clare. 2003. "Suicide Bid on CCTV May Herald New Privacy Law." *The Guardian*, January 29. www.theguardian.com/media/2003/jan/29/pressandpublishing.broadcasting.

Federal Trade Commission. 2016. "FTC Releases Annual Summary of Consumer Complaints." www.ftc.gov/news-events/press-releases/2016/03/ftc-releases-annual-summary-consumer-complaints.

Fuller, Brandon, and Paul Romer. "Urbanization as Opportunity." The World Bank Sustainable Development Network, Urban and Disaster Risk Management Department, Policy Research Working Paper 6874, Washington DC, May 2014.

Gavison, Ruth. 1980. "Privacy and the Limits of Law." *The Yale Law Journal* 89(3): 421–471.

Goffman, Erving. 1959. *The Presentation of the Self in Everyday Life*. New York: Anchor Book.

Goffman, Erving. 1963. *Behavior in Public Places*. New York: The Free Press.

Hall, Edward T. 1966. *The Hidden Dimension*. Garden City: Doubleday.

Hicken, Melanie. 2013. "Data Brokers Selling Lists of Rape Victims, AIDS Patients." *CNN*, 19 December 2013. http://money.cnn.com/2013/12/18/pf/data-broker-lists/.

Hoofnagle, Chris Jay. 2016. "Businesses Are Invading Your Privacy." *The Hill*, 27 January 2016. http://thehill.com/blogs/congress-blog/technology/267070-businesses-are-invading-your-privacy.

Laurent, Olivier. 2013. "Protecting the Right to Photograph, or Not to Be Photographed." *The New York Times*, 23 April. http://lens.blogs.nytimes.com/2013/04/23/paris-city-of-rights/.

Marmor, Andrei. 2015. "What Is the Right to Privacy?" *Philosophy and Public Affairs* 43 (1): 3–26.

Mazur, Kevin, dir. *$ellebrity*. 2012. El Segundo, CA: Gravitas Ventures, 2013. DVD.

Moore, Barrington. 1984. *Privacy. Studies in Social and Cultural History*. Armonk, New York: M. E. Sharpe.

Nagel, Thomas. 1998. "Concealment and Exposure." *Philosophy and Public Affairs* 27(1): 3–30.

Nissenbaum, Helen. 2010. *Privacy in Context. Technology, Policy, and the Integrity of Social Life*. Stanford, CA: Stanford University Press.

Rachels, James. 1975. "Why Privacy is Important." *Philosophy and Public Affairs* 4(4): 323–333.

Spacks, Patricia Meyer. 2003. *Privacy: Concealing the Eighteenth-Century Self*. Chicago, IL: The University of Chicago Press.

Véliz, Carissa. 2017. *On Privacy*. Doctoral dissertation, University of Oxford.

Welsh, Brandon C., and David Farrington. 2004. "Surveillance for Crime Prevention in Public Space: Results and Policy Changes." *Criminology and Public Policy* 3(3): 497–526.

Chapter 3

Building ivory surveillance towers

Transformations of public space in higher education

Sarah Shoemaker and Patrick Schmidt

Introduction

Much mythologizing surrounds the life of the university student. From the perspective of older adults reminiscing about their youth, one's formative years as a young adult (commonly, today, ages 18 to 22) spent in higher education are a privilege, marked by new freedoms, revelry, and experimentation. The foundation of this freedom is the space created by universities, both in the physical sense in which many students live away from home for the first time and the intellectual sense of an environment that fosters experimentation, learning from mistakes, and growth.

Given the significance of higher education as the gateway to adulthood and public life, the full impact of surveillance in society cannot be assessed without examining the effect it is working on universities. There is reason for concern. Technologies of surveillance have entered higher education's spaces and places as much as, or more than, any space in society, while around the world higher education has been subject to enormous pressure from social, economic, and political forces. What will emerge from this is only beginning to be understood.

This chapter explores the transformations that surveillance has brought to higher education, with an eye to both physical and ideational dimensions.[1] Like many traditional public fora, university campuses have been spaces for public protest. But, perhaps unlike city streets, they are also intended as spaces for the quieter work of daily intellectual development. An assessment of surveillance in this context must take stock of both the bodily control exercised by physical surveillance and the tools that work together to produce the additional layer of intellectual surveillance.

While many of the technologies used in the surveillance of higher education are the same as are employed elsewhere, in this space the consequences are profound. In order to develop this account, the first section begins with a historical turn that looks further back, perhaps, than is typical. After examining the long trajectory of higher education, in the second and third sections of the chapter we advance what we identify as a "double inversion," a two-fold transformation of the space that abruptly challenges the nature of the university as it evolved in the twentieth century. We conclude that new technologies of surveillance, the panoptic power of the institution and neoliberal paradigms of accountability are generating changes in access and behaviors within this sphere, and that these changes echo longer historical traditions of control on campuses. Our goal here is to open up a comparative examination of Western universities as public space, although we offer a caveat at the outset that there is an incredible diversity even within national traditions. This inquiry's first step requires bringing universities wholesale into the discussion of "public space."

Universities as public spaces

Thinking about the impact of surveillance on higher education is a daunting prospect, not least because of the diversity of such spaces across nations, each with a peculiar relationship and history. Accounting for the leading institutions today means recognizing that higher education in the United States has historically been distinctive, structurally and functionally, from the rest of the world (Trow 1999), and many profound differences remain despite the internationalization of the field (de Witt 2002). To add to that, universities are some of the most venerable institutions around the world, imprinting long traditions of ideas and practices in different places that can affect how surveillance is used and received. Given these complications, we focus on the institutions of Europe, the US, and Australia.[2] It is important to set out a theoretical understanding that brings this incredible variety (diverse even within individual countries) into a shared conversation. We paint with a broad brush, on a canvas in which Western institutions of higher education – for ease, we will use the imperfect term, "universities" – must be recognized as public spaces.

We understand public space as the product of two dimensions that are closely linked in any discussion of higher education. The first dimension is physical. A university is an all-encompassing entity, and includes traditional public spaces, ceremonial spaces, semi-private areas such as offices, and private spaces including living quarters. No general category of space is entirely public or private: the toilet stall of a government building is not opened to public scrutiny by virtue of its ownership. Scholars have struggled to corral the "very broad, fluid, vague, messy and dynamic paradigm" that is privacy (Bennett 2011, 514), the analytical binary to "the public," leaving exposed the fluid and multifaceted nature of space as experienced across types of places (Brown 2012) and over time (Keller 2008). We follow Parkinson's recent theorization of physical public space as linked to four distinct characteristics, which are not necessary elements of any particular space. In this framework, physical public space (1) is openly accessible; and/or (2) uses common resources; and/or (3) has common effects; and/or (4) is used for the performance of public roles (Parkinson 2012, 202).

The independence of the first three of Parkinson's factors creates a robust definition of physical public space, one that does not depend on the legal technicalities of property ownership. A seemingly private place can become public through municipal regulation or patterns of use (Kayden 2000) – a vital observation that allows us to fold together both private and public universities, even though that distinction remains relevant for narrower legal questions. Nor is it useful, under Parkinson's definition, to be delayed for long around the distinction between the daily users of the space – students, faculty, staff and administrators – and wider communities, both local and national, whose financial and social capital are invested in higher education or who look to universities to fulfill public roles. Parkinson encourages that we recognize "all affected" by a space (Parkinson 2012).

Our understanding of public space possesses a second dimension, which begins with the last element of Parkinson's framework but extends further. The "public" is not merely a physical place where individuals perform public roles, such as the offices of public trust, but the forms of communal interaction where individuals address (in speech and action) matters of public concern. In this sense, we incorporate here inspiration from the analytical categories of Hannah Arendt's "public realm" (2013) and Jürgen Habermas's "public sphere" (Habermas 1989), which describe the zone of life beyond the state itself where individuals create a collective identity. Arendt's notion, in particular, forcefully urges consideration of the social structures that enable individuals to understand and engage in dialogue about matters of public concern. Thus, when we analyze public space, we are analyzing physical spaces narrowly construed and sites demarcated by society as places where the basic concepts and terms of discourse are generated, reworked and disseminated. This definition of public space does not require that spaces afford universal and unbridled access to the public. Public space is the physical and ideational terrain in which matters of public concern are produced and contested. In this sense, the status of universities over time and in many places shows the ties between higher education and public space: changing ideas about who should have access to the public realm, the freedoms that are required there, and how public life should be organized.

Historical dynamics to the 1960s

The second dimension of public space, as laid out in the preceding section, captures the vital importance of universities during their dawn and development in Europe, even when – especially prior to the nineteenth century – many universities had a transient physical presence and were identified more by the social status of the tutors and students assembling in a locale. The early university was asked to contribute specific public goods, but social forces attempted to circumscribe its role in the public sphere. Early American universities, as preserves of social hierarchy, most often focused on the task of training clergy, with students drawn from the upper classes (Anderson 2004). Colleges did create distinctive physical spaces marked by control rather than freedom. The task of moral education required strict control, including over the living arrangements, with King's College (later Columbia University) going so far in 1763 as to remove women from residence, even though it had previously allowed the family of faculty to live there (Vine 1976). As servants of a private, corporate body, the regulation of bodies and learning were central to their existence.

The sense of universities as exclusive enclaves – driven in America by concern with moral instruction, or in Europe by the purpose of the training of state servants under the careful purview of Enlightenment monarchies (Anderson 2004) – does not alter our regard for universities as public space. Just the opposite: it makes plain how education has been tied to the constitution of the public. In the evolution of medieval universities such as Oxford and Paris, the increasing sense that scholars and students were forming a distinct place or identity within a community simultaneously generated more intense relationships with outsiders, most vividly seen in "town and gown" antagonism and riots (Catto 1984). The university was becoming a physical place with social power, and the movements in that history caused moments of renegotiation over the permeability of the public sphere and the state itself.

The dynamics of state/university relationships are complex; few works attempt to track the ups-and-downs in more than a few countries at once. Our broad strokes here aim to paint a backdrop for the developments over the past 60 years. As a general observation, prior to 1900, universities were very inaccessible spaces, especially in terms of membership, and inside, they could also subject members to extensive controls over behavior. In European universities of the early nineteenth century, the role of students in attempting revolution – think *Les Misérables* – required it. Surveillance was imperfect, especially in the largest cities, but from the historical record, we know that authorities in provincial cities minutely supervised the lives of students and subjected them to rigid discipline, especially in France, but also in Italy, Spain and other countries (Anderson 2004). Allowing for much regional and temporal variation, many practices largely extinct in our lifetimes – compulsory church attendance, mandatory academic dress, religious oaths, and strict controls on vices – were par for the course in universities, and some elements of this regime remained in place well into the twentieth century. The central control of faculty and curricula was designed into most universities of the period, as well. The fact that at Harvard, in the Puritan New England of the eighteenth century, college disciplinary practices ebbed and flowed (Moore 1976) should caution us against seeing this too monolithically. Still, while surveillance may have been uneven and imperfect, it was still an assumed part of higher education. Access to the university – and through it, positions of influence in public life – was limited by these "finishing schools" for children of the elite. Living communally in close proximity to authority figures, surveillance was the norm. As Daniel Coit Gilman, first president of The Johns Hopkins University, said, "The College implies, as a general rule, restriction rather than freedom" (Lee 2011, 67).

A fuller narrative would account for national variations in legal setting, which constructs and empowers the freedoms granted to the different publics – faculty, students and others – who engage the spaces. Even so, the broad account puts into sharp relief the transformation of higher education that culminated in the campus fervor of the 1960s. The changes began earlier as increased access and demographic diversity stoked pressure for freedom on campus (in terms of surveillance of behavior). In the United States, federal policy brought returning war veterans onto college campuses, as higher education was presented as the pathway to social mobility. Affordable college became a public priority and the doors were flung open. Civil rights and feminist movements drew many to seek fuller membership in university life. Campuses became sites of protest and focal points for the social upheaval faced by the nation as a whole. Amidst this energetic period, the US Supreme Court determined that constitutional protections, particularly freedom of speech and due process, should be applied to public American universities, creating norms deeply affecting private universities as well (Smolla 2011). European universities, over a longer timeline, trended to a convergence on rights and freedoms within academic communities. The medieval universities that were once intimately tied to the needs of the ecclesiastical hierarchy (Lawrence 1984) and had remained "pillars of religious orthodoxy, both Protestant and Catholic" through the Enlightenment (Day 2000) were set loose by waves of secularism and political liberalism.

The impetus for surveillance appeared to have been swept away. Students' energies in this new environment were not directed solely at the political crises of the age but a transformation of university life itself (Horowitz 1987). Curricula loosened, faculty claimed more independence from governing boards and administration, and students claimed – and were sometimes granted – meaningful seats at the table for the governance of the institutions. As campuses sought to become sites for the development of free citizens, universities across Europe discovered a new agenda of campus governance, student rights, curriculum and community relations (Plantan 2004).

While the second part of our definition of public space focuses attention on the evolution of universities *vis-à-vis* the public realm, the first part captures how changing ideas about the purpose of higher education have at times been rendered physical through brick and stone. Over time and campus-to-campus, the physical spaces fluctuate as more or less open, and take on different uses. To be complete, this account would need to place universities into the history of cities and national geographies. Universities in Europe grew within cities, sometimes defining a portion (or "quarter") of the urban landscape and yet curiously indistinguishable from the rest of the city. Universities in the American tradition were more wide-open spaces, yet simultaneously more distinctive and legible. The long tradition of intentional campus planning (Turner 1984) eventually spread to many other Western countries, resulting in greater numbers of suburban "garden" campuses in Europe after World War II. Formal separation from the surrounding environment is not necessarily consequential; only a small number of institutions operate as "gated communities" with total control over entry.

Depending on the motivations, efforts to surveil the university mark it out as "public" even more than its physical proximity to wider publics. Surveillance was possible in the monastic colleges of Oxford and Cambridge when students daily walked the cloisters with teachers, only its political consequences were less significant. Paradoxically, efforts to separate universities spatially in the modern era have sometimes had the effect of making them more rather than less salient to the public when campuses become bastions, often of progressive ideas (Turam 2015). Thus, the long trend-lines of privacy and surveillance in universities over time are found at the intersection of the two dimensions of public space: where the capacity for physical surveillance intersects with the desire or political will to re-align the role of the university in the public realm. How does surveillance disrupt the ideational freedom of universities, in some ways drawing the university closer to the public realm and in other ways restraining some of its denizens from public life?

During the twentieth century, surveillance became more difficult to exert on expansive, modern university campuses, and these spaces could become infused with the ideas of freedom and privacy. There is variation, of course: libraries, academic buildings, and sports facilities reflect a wide range of accessibility, from fully open to the public to restricted solely to authorized users. Some spaces, such as student housing, laboratories, and administrative offices have always been relatively more likely to require special credentials. But overall, compared to their distant pasts, in liberal democracies, higher education in the twentieth century developed a progressive narrative of inclusion and freedom both physically and ideationally. With a sense of physical distance and freedom from scrutiny, and wider access to university life, more people can envision using the university as a platform for participation in the public sphere.

The question we take up in the next two sections is: how is surveillance in higher education changing the experience and our understanding of university as a public space? Although a singular narrative about university public space is made difficult by the historical and geographic sweep of cases, it may be easier than ever to describe common tendencies, given that surveillance in higher education links strongly to managerial discourses of accountability that scholars have linked to higher education across both the Anglo-American and continental European spheres (Lorenz 2012; Pasias and Roussakis 2012). With increasing financial pressure on universities, and a rising discourse of students as consumers, universities have seen surveillance as a prime tool for streamlining the institution and managing the totality of the student experience.

Contemporary technologies of university surveillance

As viewed at the end of the 1960s, the triumphal story of Western higher education appeared to be the replacement of paternalism (universities as sites for inculcating elite values) with a new liberalism (universities as preserves of autonomy, privacy and equality for individuals seeking their understanding of the good). While some scholarship to date has discussed the impact of surveillance as an entirely novel affront to the ideals of an open, free university (Vann Heertum 2013), the availability of the paternalistic tradition – which, in the American context gives the institution *in loco parentis* responsibility for students – undermines resistance to the introduction of surveillance technologies, most commonly through video cameras/CCTV (hereafter, "cameras") and identification ("ID") cards. In their specifics, these technologies have been seen as responses to specific problems, and they offer some conveniences and benefits to the administration, security personnel, staff and students. The vital step is to assess how surveillance alters public space, understood in both the physical and ideational dimensions. Vitally, the situation is not simply the emergence of new or enhanced technologies of surveillance, but also changes in the university and its relationship to the public realm, making surveillance both a cause and an effect.

The surveillance technologies introduced into universities do more than improve surveillance; they perfect it. Unlike the medieval college bolting its doors at night, ID cards systematize access and membership, providing feedback through usage statistics, both individual and aggregate. ID cards have been present on campuses since at least the nineteenth century. However, besides serving their original purposes of granting library privileges and accounting for meals in college dining facilities, they have now (since the 2000s) been embedded with student data, magnetic strips, bar codes and proximity chips, functioning as keys and financial instruments while also producing copious data (Heldt 2012; Kennedy 2009). Nothing nefarious should be assumed about why universities have turned to ID cards. After initial investments, ID card systems reduce or eliminate the need for issuing physical keys, which are easily lost and copied. For what may be thousands of offices and laboratories, they also offer precision. Making distinctions among people is inherent to surveillance: who belongs in a space, and when? ID cards enable finer distinctions between faculty and students, among certain categories of students, and among individuals based on the time of day or calendar (i.e. academic terms).

After widespread use of CCTV by private security, cameras and video technologies made the jump to public spaces in the US, UK and much of Europe starting in the 1990s, with publicly funded systems receiving widespread public support (Ditton 2000; Clavell, Lojo, and Romero 2012; Fonio 2011; Fyfe and Bannister 1996; Webster 1996), though in some European countries, CCTV was met instead by increased regulation (Heilmann 2011). Though systematic data are unavailable, media reports and publicly posted notices show widespread adoption and often radical expansion, with security overhauls at individual universities costing millions of dollars and involving unified video systems with thousands of cameras per campus. Over the next 20 years, the quality and complexity of the systems increased, including video analytics that can track movement of individuals and allow alerts to be sent to campus security if an individual lingers too long in an area such as a cloakroom where students leave book bags that are potentially vulnerable to theft. The proliferation of IT equipment – computer labs and video projection systems in each classroom – has also helped to drive and justify the increased security.

Universities have changed in other, pronounced ways. International firms selling surveillance equipment appeal to the desire for a "comfortable environment for students and teaching staff" (Axis Communications 2015), a pitch that dovetails with the changing economics of higher education. Competition for students, including among state-sponsored universities, puts a premium on the campus experience, as do increasingly sophisticated marketing efforts and alumni fund-raising that promise a welcoming educational setting (Griffin 2006). A 1998 report offering the first survey of campus crime in the United Kingdom described the "commercial reality … to continually increase student numbers" and anticipated that crime prevention was in the process of shifting from a "peripheral" to a "legitimate and core issue" (*Times Higher Education* 2000, quoting Campbell and Bryceland 1998). In the United States, after the rape and murder of Jeanne Clery in her Lehigh University dorm room in 1986 (Gardner 2015), her parents learned that in the three years prior to their daughter's murder, 38 violent crimes had taken place on campus. Jeanne Clery's parents championed a 1990 law, later the Clery Act, which requires annual reporting of campus crime. That transparency, in turn, spurred investment in security services.

Incidents of high-profile violence on campus have played an important role in pushing institutions to adopt new technologies (Kaltefleiter 2013). Most prominently, the 2007 massacre by a graduate student at the Virginia Polytechnic Institute and State University (Virginia Tech) encouraged a security-oriented response to campus crime. The gunman's first victim, at a dormitory, was treated by the police as a case of isolated domestic violence. Virginia Tech administrators did not cancel classes or attempt to secure the rest of campus. While campus authorities met, the shooter opened fire in an academic building across campus, with a total of 32 people killed (National Public Radio 2007), at the time the deadliest mass shooting by a lone gunman in United States history (Hauser and O'Connor 2007).

With good intentions, other colleges and universities interpreted the tragedy as a call to action. Schools across the country doubled down on security cameras, introduced campus-wide alert systems through text and email, and increased security measures to identify who moves where in their spaces (Kaltefleiter 2013). While the United States is particularly criticized for its approach to firearms and gun violence, universities in many other countries have likewise sought an institutional response to rising fears of mass violence. Recording devices, identification schemes, data collection and larger police budgets often appear as rational and relatively unintrusive responses, whatever the evidence of effectiveness. Post-9/11 fears of international students and immigration, including (but not exclusively) the fear of violent extremism, have also driven the monitoring of students (Lyon 2009). At a national level, for example, in 2009 the UK began requiring non-European students and staff to carry identity cards (Marginson et al. 2010). As discussed later, an additional factor driving forms of university surveillance has been the parallel narrative about the role of mental health in mass violence. The Virginia Tech student was known to the university's counseling service, and some concerns had been raised by a faculty member as well. Though it is not clear that the shooting could have been prevented, higher education institutions have been under pressure not only to provide mental health services, but also to integrate pieces of data about students.

As mentioned earlier, differences in legal systems provide some texture to the responses in different countries. A fear of being sued for liability and negligence animates many university administrators, especially in litigation-friendly jurisdictions such as the United States. Civil liberties, especially pro-privacy regulations and jurisprudence, serve as a counterweight for some. The law does not sit unchanging, but is shaped by changing views about the proper balance between protection and liberty. In the midst of legal flux, universities have often improvised by trying to set the expectations of students and staff about the extent of video surveillance and data collection. Disclosure, while a good first step, can be couched in the same legal ambiguities that allow surveillance to go virtually unchecked. Harvard Law School specifies, for example, that cameras will not be placed in areas where there is a "reasonable expectation of privacy such as dormitory rooms, restrooms, and individual offices" (Harvard Law School 2016), where "reasonable expectation" has been the US Supreme Court's famously flexible and unpredictable standard. Stanford University's Video Surveillance System Guidelines are relatively comprehensive and include stipulations about signage (Camera in Use) and the length of time footage may be stored (no less than 32 days and no more than 365 days unless as part of a criminal investigation or court proceeding). It also makes broader promises that "Stanford does not use Video Surveillance Systems for the purpose of workplace or workforce monitoring" and, perhaps less helpfully, does "not install Video Surveillance Systems where privacy interests exceed the security value" (Stanford University 2014).

In the realm of the physical, the collective impact of these technologies undermines significantly the circa-1960s assumptions: first, that members of a university community should expect a level of bodily autonomy and non-surveillance, and second, that a university is and should be distinctive from mass publics because of its role in informing and guiding the public realm. Viewed through the long lens of history, the attempt to establish the university as a space of universal access, freed of borders and hierarchies, has now been inverted in a return to familiar assumption of universities as a place of privileges and status. Video cameras and ID cards take the historic role of physical and symbolic demarcation surrounding college campuses and make boundaries a ubiquitous aspect of internal campus space. Universities look different in some respects: faculty and administration have separated in the large modern university, perhaps altering the sources and motivations of surveillance. But the differences may not be as important as the similarities to a time past. Technologies, now with increasing sophistication, are facilitating the re-assertion of control over access and behavior.

Surveillance and discipline

In addition to the physical dimension, surveillance of public space involves an ideational dimension, by which we have meant the public sphere where ideas of collective concern are produced, tested and shared. Some of this occurs within physical spaces, but the simultaneous transformation of the university adds the vulnerabilities of a virtual world. Having re-established the centrality of access to the meaning of university physical spaces, surveillance also revives the debate about when and how to use coercive power to inculcate community values. By returning universities to spaces of limited access and behavioral control, surveillance has performed a double inversion.[3] As represented in Figure 3.1, the move that took exclusive physical spaces into a period of openness and then back into a controlled space ordained by administrative authorities has its parallel in the ideational transformation of higher education from a setting for inculcation to a liberal community and then again to a place that emphasizes discipline.

The re-alignment of surveillance to paternalistic goals speaks to the changing role of the university in the public realm, and it corresponds to the concern with the self-discipline that it engenders among students and faculty. What Foucault (1977) envisioned, when translating the image of Jeremy Bentham's Panopticon to the technologies of control in contemporary life, matches the direction of changes in the learning environment allowed by surveillance and the rise of virtual life in higher education. The beauty of a panopticon – from the perspective of its administrators – is its efficiency. The inmates of Bentham's prison do not need to be observed continuously; they must assume the presence of oversight. Even if there are natural limitations on the ways schools watch students (in law, norms or technology), the university produces a panopticon that shapes both the physical behavior – including protest and other forms of bodily expression – and ideational or intellectual freedom of the community's members.

The fundamental lesson of the panopticon has been validated by emerging studies showing how surveillance affects the ideational realm. Reminding people that they could be watched online changes what many are willing to

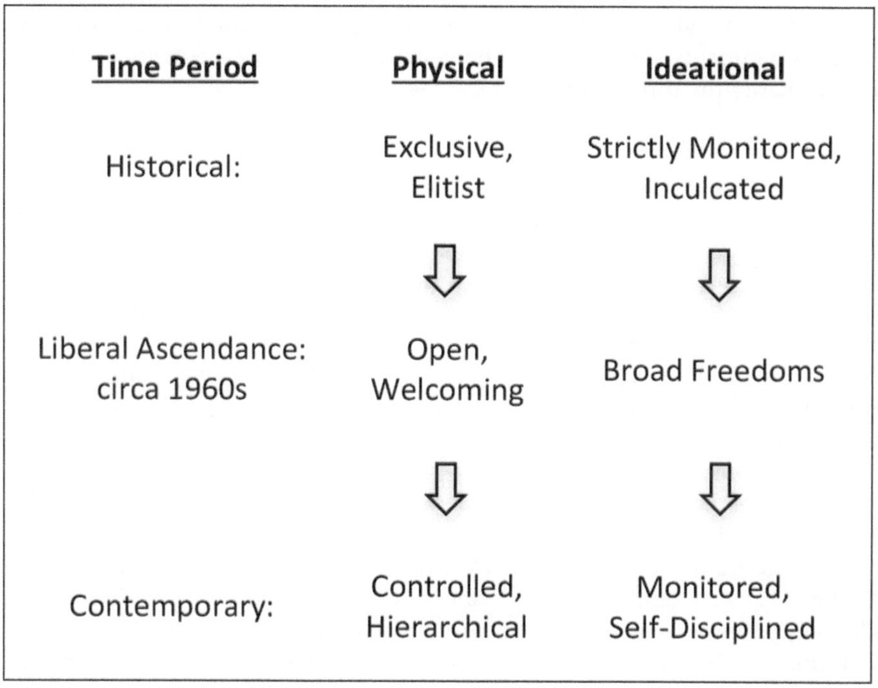

Figure 3.1 The Double Inversion of University Surveillance

discuss (Stoycheff 2016). A so-called "spiral of silence" hushes opinions among those with low to moderate certainty (Matthes, Rios Morrison, and Schemer 2010) as people seek to appear to align or nearly align with what they perceive as the most widely held belief (Stoycheff 2016). The presence of cameras changes behavior for some more than others (Rompay, Vonk, and Fransen 2009). The aspiration of universities as public spaces characterized by wide-open, critical discourse is among the goals most vulnerable to the pathologies of a surveilled environment.

Bentham predicted success without accomplishing total surveillance, and even if the future of surveillance promises deeper reaches and more creative means of surveillance, the range and increasing creativity of surveillance long ago surpassed the point sufficient to bring Foucault's concern to bear. Campus surveillance systems, through evolving technologies and practices, lace together the pervasive video record and data collection. Once digitized, personal records can be connected to physical surveillance systems, with increasing potential for understanding the attitudes and activities of community members. On any given day, a student might walk from their residence hall to the library with cameras capturing their image in the elevator and as they exit and enter the buildings. Their first task at the library might be to check the online coursework system, say Blackboard or Angel, where their professors have posted assignments and reading. The system will log their activity, displaying for the faculty the time and IP address when they accessed the day's assignment. Their reactions to the readings, submitted online, can be recorded and at many universities will be permanently archived. The student might then use their ID card to check out books that will help them contribute to an online class discussion, or, more likely, use their university credentials to access an academic source in an online database to inform their answer. If they attend a traditional class, one or more students may have been authorized to record the class in order to accommodate their disabilities, if the institution has a policy limiting the use of recording devices in classrooms and all.

Next, the student might use the credentials on their ID card again to reserve a room on campus for the club to meet, later checking social media. Should the student decide to skip a class, their professor may log their absence, initiating an automatic email to their academic advisor, on an email account provided by the university. The advisor may log the interaction on a system designed to identify students at risk of attrition that also contains notes on the student from their residence hall director. One provider of such software (Campus Labs), selling its "Beacon" software, hints at the image of a lighthouse, allowing university officials to see through a fog of information about student performance. As suggested earlier, such systems may also connect university staff with information from college mental health services. When matriculating, the student signed a consent release giving the school permission to enter their information into this software. The potential of information systems to paint a nuanced portrait of a student over time undermines privacies that were once protected by the sheer inefficiency of collecting information all in one place.

In the vision of a panopticon, if students simply assume that wayward behavior will be caught, then surveillance would not need to be total in order to produce the desired behavior. Nevertheless, a bevy of commercial products specifically promise to nudge students toward desired habits. Some tools make themselves explicit. A small school in the American south, Lynn University, issues iPads to students that sound an alarm and alert a student's academic advisor if the machine is not in class (Kelly 2015). Systems that coordinate information offer significant benefits to students and institutions, particularly by boosting student retention and program completion. The intention behind the technology is to connect struggling students to the resources that can help them sooner and more effectively; of course, improved retention also benefits colleges, because students who complete a degree pay more tuition than those who do not. The rapid development of the virtual dimension of higher education has outstripped reflection and legal developments. Some boundaries in the use of all this data are established by law. Other limits can be technological, either practically speaking because the technology is not advanced enough to coalesce all information, or by design so as to only give access to certain staff. University librarians in the United States have been a source of resistance, with professional efforts to resist surveillance and federal terrorism investigations by programming systems to delete user records after library materials are returned.

These effects of routine surveillance are intensified by digital memory. Unlike the surveillance of the pre-modern university, which could be brutally effective, the memory of "youthful indiscretion" was limited to a few individuals and could fade over time. Accountability is important, but discipline is less dangerous when the punishment is not permanent. Today, perfect, infinitely reproducible records mean that not only do failures and mistakes not disappear, they are increasingly accessible to others long after one has left that community (Mayer-Schönberger 2011). Europe has moved much further at this point in recognizing that a right to be forgotten is essential to the effective value of privacy (Jones 2016), but it will be a long time before we can assess the chilling effect that digital memory has on the culture of learning in higher education.

Even if new legal boundaries ultimately resist the panopticon, individuals will now depend on the grace of the state to maintain the university as a space of access, autonomy and public-spirited engagement. Too much faith would be misplaced. In a 2016 incident at Emory University in Atlanta, Georgia, students awoke to "Trump 2016", and other short messages supporting then-candidate Trump, written in chalk on campus sidewalks. Such slogans had taken on the air of racial epithets, owing to Donald Trump's inflammatory rhetoric. Offended students took it up with the administration, and University President James W. Wagner indicated that the school would review surveillance footage and investigate the incident as student conduct violations (Singhal 2016). Outrage at the students' outrage itself spurred competing calls for freedom of speech. While the university made clear that it would not review surveillance footage to punish the authors, the administration did publically recognize the original student concerns about maintaining a safe and respectful campus environment, a stance that was met with continuing criticism from outside observers.

Setting aside the debate about "political correctness" and sensitivity that was triggered by this case of anonymous political speech in a public campus space (albeit a private university), the incident shows the manipulability of the values to which surveillance is put into service. The enforcement of legitimate regulations about hate speech quickly elides into the management of dissent and conflict, where the desire for "peace" and "community" reflects the reputational concerns of an institution that must treat students as consumers. Likewise, university administrators have overlooked the tensions or outright contradictions in their cooperation with national security agencies who have sought to infiltrate higher education (Golden 2017). Most of all, the greater challenge for observers of trends in higher education involves unwinding the connection between such surveillance and the emergent cultures that have grown out of the "audit culture" that has transformed higher education worldwide (Shore 2008; Suspitsyna 2010). As discussed earlier, after universities expanded in the twentieth century to widen access and pursue the goal of social mobility, the counter-movement of the past several decades has attempted to reallocate – pass on – the resulting costs. Families in the United States have long paid high levels of tuition for private university educations, and tuition levels for public universities have risen sharply, though they still benefit from some public subsidy. Fees have been introduced or are increasing in many European countries as governments likewise seek to reduce their subsidies and investments. The squeeze has meant that universities are under greater pressure to prove the economic value of their services. Data collection allows institutions to allocate resources more efficiently, and to demonstrate their value, but with the effect of transforming curricula and courses, as institutions and "consumers" alike seek to extract more of that value. New, surveillance-ready educational technologies, like ID cards for the physical dimension, support this environment with data on individual and institutional performance. Student performance is about the good of the student and the health of the institution: graduates who can earn a salary are able to pay their tuition and contribute to the reputation of the institution. Campus safety serves that same end. All told, the pressures point toward management, and if students learn neoliberal self-discipline, that is a welcome outcome as well.

In the nature of things, it is more difficult to demonstrate the effect of surveillance on the ideational role of the university than it is to observe the widespread manipulation of the physical environment. In the double inversion that we identify here, the latter contributes in some measure to the former by creating the awareness of surveillance and chilling the environment for students and faculty. We will need greater distance to see the consequences of ubiquitous surveillance in the realm of intellectual inquiry, but the warning signs and threats to the public realm are present.

Conclusion

The university is a public space, because of both the nature of its physical spaces and its ideational connection to the public realm. Its status as public space is established not simply by its physical relationship to the community, though that is very clear in many places. Nor is it established by public ownership, which is murky even as state institutions today continue to invest heavily in higher education. Visiting campuses across multiple countries brings one to a wealth of spaces, with various forms of demarcation and a multiplicity of functions. At the heart, universities are public spaces because of how they present themselves to mass publics, but even more because they are sites of contest: inquiry, debate and a struggle for access, as individuals seek to use education and expertise as the pathway to control of the public realm. The university is a space where the meaning of and ownership over the common good is at stake.

That contest can flow toward access and it can flow toward restraint. The past century ushered in a sea-change in how that space was used, with the introduction of new expectations that the space would offer unprecedented access and freedom. The rise of surveillance technologies puts those changes into doubt. Through the physical surveillance technologies of ID cards and video cameras, the spaces of higher education are now more closely observed than ever, and the urgency behind the manipulation of those spaces has been fed by 20 years of fear of violent crime and terrorism. At the same time, the simultaneous rise of data collection, preservation and manipulation has made the university subject to scrutiny and economic crosswinds. The results have been steps within the university environment that are justified by reference to the logics of economics and the university's consumers. Community values of order militate against the deviant, resulting in a tighter, less open-textured intellectual enterprise. With the benefit of a long historical perspective, we view these transformations as a double inversion, in which two trends have been reversed and the liberal ideals of the university have been subverted by the new priorities.

Only with the benefit of much more hindsight can we assess fully the consequences of this transformation. The idealized purpose of the university in the twentieth century, and one that continues to inspire, is to prepare all students to not only enter but also share in the project of forming the meaning and direction of the public realm. The public spaces of higher education provide the setting in which students model and test out their vision for the wider community in the future. When universities impose panoptic discipline on students, do they implicitly train the next generation to not only expect but also favor that form of power? When universities implement widespread, efficient systems of surveillance with so little student input or oversight, and little obvious recourse should they object, do they miss a crucial opportunity to engage youth in a discussion of surveillance, power and privacy? Ultimately, the challenge will be to fling open the doors of the ivory tower to let in just such discussions and to create educational spaces that can produce adults who are free enough to question the conditions of the surveillance society being created for the public.

Acknowledgements

The authors would like to thank Brian Rosenberg, Terry Boychuk, the editors, and an anonymous reviewer for their thoughtful contributions to this chapter.

Notes

1 By physical, we mean "movement of bodies through space". By "ideational" we intend a broad understanding of processes by which ideas and the production of ideas are influenced, including ideas about the constitution of the public itself, and for purposes here we use it interchangeably with "intellectual".

2 Developments in many places, such as Asia or the former colonies of the British Empire (which exported its educational system broadly), may share some commonalities with our account but lie outside the scope of our examination.

3 In our choice of terms we acknowledge a debt to Van Heertum (2013) but we extend his analysis in two ways, in what might well be described (more playfully) as a *double-double* inversion: first by locating the contemporary transformation as the outcome of a longer historical struggle that is reverting in at least some respects to a prior set of assumptions, and second by identifying the inversion as occurring along two dimensions.

References

Anderson, R. D. 2004. *European Universities from the Enlightenment to 1914*. Oxford: Oxford University Press.

Arendt, Hannah. 2013. *The Human Condition*. 2nd ed. University of Chicago Press.

Axis Communications. 2015. "Safe Education at Karaganda State Medical University." www.axis.com/files/success_stories/ss_edu_ksmu_64597_en_1508_lo.pdf.

Bennett, Colin J. 2011. "In Further Defence of Privacy ..." *Surveillance & Society* 8(4): 513–516.

Brown, Lance Jay. 2012. "Public Space Then and in the Future." In *Beyond Zuccotti Park: Freedom of Assembly and the Occupation of Public Space*, edited by Ron Shiffman, Rick Bell, Lance Jay Brown, and Lynne Elizabeth, 236–253. Oakland, CA: New Village Press.

Campbell, Kenneth, and Charles Bryceland. 1998. *Policing the Campus: Providing a Safe and Secure Environment*. London: Police Research Group.

Catto, J. I. 1984. "Citizens, Scholars and Masters." In *The History of the University of Oxford*, edited by J. I. Catto, 151–192. Oxford: Oxford University Press.

Clavell, Gemma, Gladon, Lohitzune Zuloaga Lojo, and Armando Romero. 2012. "CCTV in Spain: An Empirical Account of the Deployment of Video-Surveillance in a Southern-European Country." *Information Polity: The International Journal of Government & Democracy in the Information Age* 17(1): 57–68.

Day, Charles R. 2000. "Higher Education." In *Encyclopedia of European Social History*, edited by Peter N. Stearns. New York: Scribners & Sons.

Ditton, Jason. 2000. "Crime and the City: Public Attitudes towards Open-Street CCTV in Glasgow." *The British Journal of Criminology* 40(4): 692–709.

Fonio, Chiara. 2011. "The Silent Growth of Video Surveillance in Italy." *Information Polity: The International Journal of Government & Democracy in the Information Age* 16(4): 379–388.

Foucault, Michel. 1977. *Discipline and Punish: The Birth of the Prison*. 1st American ed. New York: Pantheon Books.

Fyfe, Nicholas R., and Jon Bannister. 1996. "City Watching: Closed Circuit Television Surveillance in Public Spaces." *Area* 28(1): 37–46.

Gardner, Lee. 2015. "25 Years Later, Has Clery Made Campuses Safer?" *The Chronicle of Higher Education*, March 9. http://chronicle.com/article/25-Years-Later-Has-Clery/228305.

Golden, Daniel. 2017. *Spy Schools: How the CIA, FBI, and Foreign Intelligence Secretly Exploit America's Universities*. New York: Henry Holt and Co.

Griffin, Owen R. 2006. "Confronting the Evolving Safety and Security Challenge at Colleges and Universities." *University of New Hampshire Law Review* 5(3).

Habermas, Jürgen. 1989. *The Structural Transformation of the Public Sphere: An Inquiry into a Category of Bourgeois Society*. Cambridge, MA: MIT Press.

Harvard Law School. 2016. "Video Security Policies." Harvard Law School. http://hls.harvard.edu/dept/facilities/video-security-policy/.

Hauser, Christine, and Anahad O'Connor. 2007. "Virginia Tech Shooting Leaves 33 Dead." *The New York Times*, April 16. www.nytimes.com/2007/04/16/us/16cnd-shooting.html.

Heilmann, Eric. 2011. "Video Surveillance and Security Policy in France: From Regulation to Widespread Acceptance." *Information Polity: The International Journal of Government & Democracy in the Information Age* 16(4): 369–377.

Heldt, Diane. 2012. "Colleges Expanding Use of All-In-One Student ID Cards." *Community College Week* 24(25): 10–10.

Horowitz, Helen Lefkowitz. 1987. *Campus Life: Undergraduate Cultures from the End of the Eighteenth Century to the Present.* Chicago, IL: University of Chicago Press.

Jones, Meg Leta. 2016. *Ctrl + Z: The Right to Be Forgotten.* New York, NY: NYU Press.

Kaltefleiter, Caroline K. 2013. "We Are All Hokies." In *Policing the Campus: Academic Repression, Surveillance, and the Occupy Movement*, edited by Anthony J. Nocella I and David Gabbard, 105–122. New York: Peter Lang Publishing Inc.

Kayden, Jerold S. 2000. *Privately Owned Public Space: The New York City Experience.* New York, NY: Department of City Planning and Municipal Art Society of New York.

Keller, Lisa. 2008. *Triumph of Order: Democracy & Public Space in New York and London.* New York, NY: Columbia University Press.

Kelly, Rhea. 2015. "Lynn University to Track Student Attendance by iPad Location." *Campus Technology*, March 9. https://campustechnology.com/articles/2015/03/09/lynn-university-to-track-student-attendance-by-ipad-location.aspx.

Kennedy, Mike. 2009. "Keyless Access." *American School & University* 81(9): 32–34.

Lawrence, C. H. 1984. "The University in State and Church." In *The History of the University of Oxford: The Early Oxford Schools*, edited by J. I. Catto, 1:97–150. Oxford: Oxford University Press.

Lee, Philip. 2011. "The Curious Life of *In Loco Parentis* at American Universities." *Higher Education in Review* 8: 65–90.

Lorenz, Chris. 2012. "If You're So Smart, Why Are You under Surveillance? Universities, Neoliberalism, and New Public Management." *Critical Inquiry* 38(3): 599–629. doi:10.1086/664553.

Lyon, David. 2009. *Identifying Citizens: ID Cards as Surveillance.* Cambridge: Polity.

Marginson, Simon, Chris Nyland, Erlenawati Sawir, and Helen Forbes-Mewett. 2010. *International Student Security.* Cambridge: Cambridge University Press.

Matthes, Jörg, Kimberly Rios Morrison, and Christian Schemer. 2010. "A Spiral of Silence for Some: Attitude Certainty and the Expression of Political Minority Opinions." *Communication Research* 37(6): 774–800. doi:10.1177/0093650210362685.

Mayer-Schönberger, Viktor. 2011. *Delete: The Virtue of Forgetting in the Digital Age.* Princeton University Press.

Moore, Kathryn McDaniel. 1976. "Freedom and Constraint in Eighteenth Century Harvard." *The Journal of Higher Education* 47(6): 649–659.

National Public Radio. 2007. "Timeline: How the Virginia Tech Shootings Unfolded." NPR.org, April 17. www.npr.org/templates/story/story.php?storyId=9636137.

Parkinson, John R. 2012. "Conclusions and Implications." In *Democracy and Public Space: The Physical Sites of Democratic Performance*, 191–222. Oxford: Oxford University Press.

Pasias, George, and Yannis Roussakis. 2012. "'Who Marks the Bench?'A Critical Review of the Neo-European 'Paradigm Shift' through Higher Education Policies and Discourses." *The Journal for Critical Education Policy Studies* 10(1): 127–141.

Plantan, Frank. 2004. "The University as a Site of Citizenship." In *The University as Res Publica*, edited by Sjur Bergan, 83–128. Strasbourg: Council of Europe Publishing.

Rompay, Thomas J. L. van, Dorette J. Vonk, and Marieke L. Fransen. 2009. "The Eye of the Camera: Effects of Security Cameras on pro-Social Behavior." *Environment and Behavior* 41(1): 60–74.

Shore, Cris. 2008. "Audit Culture and Illiberal Governance: Universities and the Politics of Accountability." *Anthropological Theory* 8(3): 278–298.

Times Higher Education (THE). 2000. "Should an Ivory Tower Turn Fortress?" March 3, 2000. www.timeshighereducation.com/news/should-an-ivory-tower-turn-fortress/150449.article.

Singhal, Jesse. 2016. "Emory University: We Will Use Security Cameras to Track Down and Possibly Punish Students Who Chalked Their Support for Trump." *Daily Intelligencer*. March 22. http://nymag.com/daily/intelligencer/2016/03/emory-u-to-track-down-trump-supporting-chalkers.html.

Smolla, Rodney A. 2011. *The Constitution Goes to College: Five Constitutional Ideas that Have Shaped the American University*. New York: NYU Press.

Stanford University. 2014. "Stanford University Video Surveillance System Guidelines." Stanford University, May 1. https://police.stanford.edu/pdf/vssguidelines.pdf.

Stoycheff, Elizabeth. 2016. "Under Surveillance: Examining Facebook's Spiral of Silence Effects in the Wake of NSA Internet Monitoring." *Journalism & Mass Communication Quarterly* 93(2): 296–311.

Suspitsyna, Tatiana. 2010. "Accountability in American Education as a Rhetoric and a Technology of Governmentality." *Journal of Education Policy* 25(5): 567–586. https://doi.org/10.1080/02680930903548411.

Times Higher Education. 2000. "Should an Ivory Tower Turn Fortress?" *Times Higher Education*, March 3. www.timeshighereducation.com/news/should-an-ivory-tower-turn-fortress/150449.article.

Trow, Martin. 1999. "From Mass Higher Education to Universal Access: The American Advantage." *Minerva* 37(4): 303–328.

Turam, Berna. 2015. *Gaining Freedoms: Claiming Space in Istanbul and Berlin*. Stanford, CA: Stanford University Press.

Turner, Paul Venable. 1984. *Campus: An American Planning Tradition*. Cambridge, MA: MIT Press.

Van Heertum, Richard. 2013. "Cameras and ID Card Swipes: Disappearing Privacy and the Cultivation of the Virtual Self." In *Policing the Campus: Academic Repression, Surveillance, and the Occupy Movement*, edited by Anthony J. Nocella II and David Gabbard, 67–77. New York: Peter Lang Publishing Inc.

Vine, Phyllis. 1976. "The Social Function of Eighteenth-Century Higher Education." *History of Education Quarterly* 16(4): 409–424.

Webster, William C. W. R. 1996. "Closed Circuit Television and Governance: The Eve of a Surveillance Age." *Information Infrastructure & Policy* 5(4): 253–263.

Witt, Hans de. 2002. *Internationalization of Higher Education in the United States of America and Europe: A Historical, Comparative, and Conceptual Analysis*. Charlotte, NC: Information Age Publishing.

Chapter 4

The changing nature of public space in São Paulo

A taxonomic approach

Anthony Boanada-Fuchs

Introduction

Public spaces are the backbone of cities and the way we experience urbanity. At the same time, the idea of public space is considerably larger than its mere spatial dimension and includes important political, social, cultural, and legal dimensions. From such an understanding, public spaces may also be perceived as mirrors of society and its cities, embodying the balance between publicness and privacy, freedom and restrictions, surveillance and security. Due to this multi-dimensional nature, public space represents a great academic lens to study urban change and its underlying forces. Research can be quite challenging given that such dynamic relationships create a lot of uncertainty and confusion about the definition, implications, and measurement of public space. "What do we mean by public space?", "Why is public space important to us?", and "How can public space contribute to a more sustainable society?" are just three fundamental, yet not easily answered, questions.

This chapter uses these questions as a starting point to venture into the complexity of public spaces, specifically to unfold the ways in which past and more recent economic, spatial, and societal transformations have changed the nature of public spaces, and how the altered nature of these spaces challenges established discourses. In the following sections, I retrace the changes in cities and societies in Europe and the United States and then discuss the broad spatial changes in Brazil and the city of São Paulo. While the US developments are well-documented and serve to introduce the reader to the general parameters of spatial transformations and their underlying causes, this chapter helps to cast light on a rather untold story (at least to non-Portuguese speakers), of the history and trajectory of changes to public spaces in Brazil, with special reference to São Paulo.

In the context of an ongoing hybridization of public space that juxtaposes publicness and privacy, more nuanced analytical tools are needed to capture the diverse nature of spaces. I propose a taxonomy that allows us to analyze traditional forms of public spaces (town squares, streets, parks) as well as new forms (corporate plazas, shopping malls, gated communities) equally. The taxonomy is tested in a case study of São Paulo, a city known for its security infrastructure ("city of walls") and crime-induced segregation that accelerated the demise of traditional public spaces and a proliferation of its hybrid substitutes. São Paulo is therefore a *cas d'excellence* to study the changing forms of public space and how public space links to control and surveillance. The public spaces of the city have undergone a considerable hybridization with an increasing privatization of ownership and management, functional changes, and establishment of security measures. The proposed taxonomy, with four categories and a total of 40 variables, is capable of capturing the diversity of modern publicly accessible spaces while helping to identify the differences between the various types.

Public space is dead – long live public space: the historic transformation of public space

Any meaningful discussion of public space needs to equally account for the role of publicness and privacy, as the dialectic understanding of both shape our understanding and perception of space. Indeed, when looking at the historical evolution of public space, major changes in its use appear to be correlated with broader changes in the perception of privacy. Within the limitations of historical research, it is possible to show that changes in privacy have affected the way public life has spatially unfolded – which, in turn, has impacted how privacy was perceived. Thus, there is a complex dialectic relation between space and society that can be tentatively retraced in the historical evolution of the perception of privacy and publicness and physical changes in public space.

From a Western and etymological perspective, the distinction between public and private can be traced back to the Roman Empire and different legal realities. The Roman law differentiated between *publicus* and *privates* (withdrawn from public life). The *res publicae* (government-owned) was only one of a total of three legal domains, next to *res communes omnium* (domain of society) and *res divini juris* (domain of religion, Harteveld 2006, 38f). Privacy in Roman times was linked to the typology of the household (*domus*) and less strictly perceived than our modern meaning. Still, this domain guaranteed certain (but not complete) liberties from public scrutiny and control.

In the English language, the terms "public" and "private" emerged in the late 15th and 16th century, respectively, but it took until the 18th century to gain their modern meaning as "open to the scrutiny of anyone" and "a sheltered region of life defined by one's family and friends" (Sennett 1992, 16). Privacy was equated with the spaces of the domestic life and the family, and thus much narrower than the Roman *domus* or Greek *oikos*. Public life consisted of activities beyond them.

While most people nowadays would agree with such definitions, Richard Sennett eloquently showed how this constellation slowly changed over the last two centuries. Socio-spatial transformation triggered by the industrial revolution solidified the private sphere while increasing emotional distance to the public, making it ever more anonymous. Three forces led to the erosion of public life as previously known to society. The increasing commodification of public life was made possible by newly generated capital seeking new investment possibilities, the reformulation of secularism and the emergence of credo substitutes to fill the void, and the fact that public life did not collapse despite its major demise in the Ancient Regime (Sennett 1992, 19). The common effect of these changes is that cities became places of accelerated change, without offering the emotional and social backing of established institutions.

The space of privacy was seen as a last stronghold of protection that allowed an escape from the formalities and threats of society (Sennett 1992, 3f). Urbanites of the early 20th century reduced their presence, or at least interaction, in public space. The erosion of public life and ossification of privacy was reinforced by the development of an additional interface as people started carving out a selective (or even different) public image of themselves. Self-representation allowed the control of what is shown of someone's personality and status to the public. The *theatrum mundi* has evolved into a *spectaculum mundi* – while the former can be characterized by a "union of aesthetics and social reality" (Sennett 1992, 313), the latter celebrates the visual over interaction and imposes a clear distinction between privacy and publicness. Such psychology encourages the development of private cultures and individualized societies that ultimately have eliminated the societal need for traditional public space. Interestingly, in the same period, new hybrid forms of public spaces emerged in the city: arcades and department stores were developed as the new public spaces of the urban elites but were, technically, privately owned and managed. These spaces offered luxurious and protected environments and represented a strong contrast to the bustling and polluted streets of the city.

The declining societal importance of tradition public space continued throughout the 20th century but was accelerated by major spatial transformations. Contrary to the privately owned public spaces of the early 20th century which were still embedded in a network of public spaces, large parts of the urbanization undertaken post-World War II created complete and disconnected substitutes (for a more detailed discussion, see Cohen 1996).

These important geographic but also socio-economic transformations can be very clearly traced in the United States, where the GI-Bill (1944) and the Highway Act (1956) institutionally enabled individual mobility, mass production of commercial goods, and large-scale road infrastructure projects (Harteveld and Fuchs 2007, 4). The North American developments preceded most developments in other parts of the world but, as we shall see, share great similarities with developments in other regions, such as Latin America. It is, therefore, worthwhile to explain the spatial changes in the United States in more detail, as these are very well documented.

From the 1950s onwards, a broad process of suburbanization took place in the largest cities of the country. Families with the necessary financial means and the readiness to commute large distances decided to live far from work and leisure places but also the nuisances of the city (noise, pollution, crime, etc.). Residential suburban communities offered simulated archaic neighborhood experiences based on the idea of garden cities.[1] These places already featured the main attributes of gated communities: security, privacy, and a high-level of services (see Frantz 2006, 70). Perimeter fences and gates ensured that only a narrowly defined public was able to access and enjoy the benefits of such gated communities.

While the idea of gating is hardly new (examples can be found in socio-historical diverse contexts such as Roman and Greek Antiquity, 19th century London, Berlin and Paris, 20th century New York, and even Latin America),[2] the historic examples remained isolated episodes (Morris 1972). Since the end of World War II, gating has become a common (and, in some parts of the world, the dominant) feature of urban development. The success of gated housing estates can be seen in the creation of an artificial community based on shared characteristics often perceived as important (income, status, lifestyle, etc., Blakely and Snyder 1997). In contrast to large and anonymous housing developments in the city, suburban residential developments provide single-family houses in well-defined public (semi-public, semi-private) and private spaces. Such transition does not only create a spatial buffer between the privacy of home and the publicness of the city but also conveys ownership and surveillance (see Newman's defensible place theory; 1972, 1996).

The popularity of gated residential communities in the United States stimulated an increasing demographic and fiscal shift towards the urban periphery – a process that soon was accompanied by services and facilities. Shopping centers and office parks offering places of leisure and work represented new suburban forms of centrality. Like the residential solutions and their simulation of community, office parks and shopping malls create an illusion of public space. The levels of control not only determined who can visit such places (directly through controlled access or indirectly by car-based design), but also included an increasing set of environmental conditions. The unpredictable "urban ennui" was avoided by simulating and controlling lighting, climate, nature and keeping out noise, pollution, and often unwanted societal effects (Banerjee 2001, 13; Harteveld 2006, 36). How much public is left in these places? The ownership and management of such spaces is de facto private. Publicly accessible private spaces offer fragmented micro-cosmoses of social reality by welcoming a well-defined public of a specific socio-economic reality. In this regard, the "privatization of space and the threat of social fragmentation pose a serious threat for the future of the city" (Madanipour 2003, 228).

The American story of public space transformations has been retold in many countries. Around the world, successful places of public life are often strange hybrid forms (Kohn 2004, 11), combining private ownership and/or maintenance, restricted access, a plethora of security technologies, and an aura of commercialization. Technological possibilities, such as advances in information and communication technologies and surveillance technologies, have redefined our very understanding of public space, eroded our need for it, and even enabled substitute forms to physical social exchange. Given these developments, we may wonder whether public space has been entirely dematerialized and we are witnessing the "end of public space?" (Sorkin 1992; Mitchell 1995). One look at the activity level in shopping centers around the world suffices to realize that space and physical presence is still relevant for social life. It appears that it is just the quality and requirement of publicness of space that has changed (Low and Smith 2006). In the next section, I address the academic debate spawned in response to this changed reality.

Public space as an academic lens

The idea of public space is strongly connected with visually observed physical space. The former section of this chapter discussed, in order of importance, the underlying mechanisms that lead to the continuous demise of public space in terms of use, symbolic meaning, and spatial quality. The transformations of public space are largely reflected in the academic debate. Since the 1960s,[3] discussions about public space have considerably diversified. For this chapter, two debates are of relevance: the first is on the public sphere, mostly led by political scientists, and the second is on physical public places, and is largely influenced by planners, architects, and community activists.

Political scientists largely discussed the role of public space in society, perceiving it as a basic requirement to develop civility and democracy (Mitchell 1995, 115). Public space is seen at the same time as political space – the place of the *res publica* (in contrast to the *oikos)* – as well as social – the place of sociability (*Gesellschaft*, in contrast to domesticity, see the distinction in Weintraub 1997). Interactions and the presence of others are essential factors of public space (Young 1990; Kohn 2004). The public manifests in moments when people lay down their differences and start interacting with one another, an argument central to Hannah Arendt's "human condition" and also the work of Jürgen Habermas on the public sphere (*Öffentlichkeit*).

The (idealized image of the) polis and agora are central ideas in Arendt's work stressing that in the public realm humans are freed from their oppression and can act politically (*zóon politikón*, often incorrectly referred to as social, see Fuchs et al. 2009). Unlike the other dimensions of human life that might be done in solitude, such activity requires the other in order to be carried out. This action (*handeln*) consists of different levels of exchange including seeing, being seen and, most importantly for Arendt, speaking and being heard. We may thus distinguish between the public sphere/public realm as the *space* of publicness from public space as the *place* where such exchanges take place. Public space should enable visibility, accessibility, and interaction of people (Madden 2010, 189). In other words, a public space that is not accessible to all, threatens the true nature of human beings in acting politically. However, the discourse is pre-occupied with the idea(l) of public space and also remains largely silent on how to achieve this aim (see Mitchell 1995, 117).

The academic debate on the physical nature of public space has shared the concerns of political scientists but linked these with actual places. In many areas of the world, post-war urbanization has undermined and displaced public places, leading to the proliferation of gated communities and privatization of urban environments. In a very practical analysis, the protagonists of the discourse tried to identify the factors needed to make declining places great again and/or identify the underlying reasons for successful public spaces. For example, the Project for Public Spaces (PPS), a non-profit organization following the work of William H. Whyte, has developed several practical tools for implementing urban designs. It distinguishes four dimensions of public spaces: sociability, uses/activities, access/linkages, and comfort/image (PPS 2000). Essentially, success is seen as an achievement of the vote by feet; places with large numbers of people are considered successful. These investigations are limited to the traditional forms of public places such as squares, streets, and parks. In reality, however, many of these places have been results of neoliberal urban renewal campaigns, have a commercialized environment, and are, legally speaking, privately owned and managed spaces.

Towards a taxonomy of public space

The ambivalence of publicly accessible spaces gave new impulses to the theoretical and practical debate. More recently, some scholars have treated publicness not as a given but as an analytical category. Kohn proposed to answer the complexity of public space by developing a cluster concept of categories (Kohn 2004, 11). His categories included ownership next to accessibility, and intersubjectivity. Marcuse developed six different categories of publicness based on different uses, functions, and ownerships (Marcuse 2005). Carmona, with an additional concern for function and perception, argued for a public–private continuum and developed a system of four groups and 20 different types of spaces (that also included shopping malls, Carmona 2010, 2015; see also for a different proposal, Lopes, Cruz, and Pinho 2012). Lastly, Varna and Tiesdall have proposed five meta-categories: ownership, control, civility, physical configuration, and animation (Varna and Tiesdell 2010, 576). While many categories are shared among these frameworks, there remain enough differences to prevent the development of a comprehensive and integrated model applicable to all kinds of spaces and geographies.

A promising way forward for developing a comprehensive framework is the one proposed by Németh and Schmidt (2007, 2011). In order to respect the granted chapter length, I shall concentrate on their work to discuss the new generation of theoretical public space models (see Table 4.1).

The authors use three components to distinguish between different levels of publicness: ownership, management, and uses/users. Using differences in ownership as entry point and focusing on the issue of management, the authors attempt to assess the different effects on use and users. Concentrating on a broader understanding of place management captures the most essential physical characteristics (access, infrastructure, use, design) but also the intangible aspects of place (image, laws and surveillance). Such focus also shifts the attention from the degree of publicness to assessing the level of control. In other words, public or private ownership and management may influence the characteristics and uses of a place less than the way control is exerted.

Based on fieldwork in New York, the authors developed 20 variables with direct feedback from local experts. The management variables were organized into four categories, equally accounting for the effects of control/discouraging use and encouraging use (Németh and Schmidt 2011). These effects can occur in the categories of design/image, access/territoriality, surveillance/policing, and law/rules. For example, the variable "designing for appropriate use" has a controlling effect on the use of a place, while offering a "diversity in seating types" increases possibilities and encourages use. This framework has proven to be operational for the analysis of private and public squares but also public parks, albeit almost exclusively in the United States.

The main advantage of the discussed models is their capacity to reduce the diversity of publicly accessible spaces into one single framework, enabling comparison among similar and between very different spaces. This can generate important insights on the different characteristics of publicly accessible spaces and explain why certain places "work" (in terms of users, functions, etc.) better than others. Furthermore, the discussion of control moves the analysis closer to the concerns of this volume. Changes in the perception and use of publicly accessible spaces are driven by a complex constellation of the need for privacy and security. In the post 9/11 landscape of the United States, Németh and Schmidt (2011, 20) were able to show that particularly privately owned and managed spaces deploy considerable measures to control and survey user behavior.

With the aim of developing a taxonomic understanding of public space, I would like to raise three limitations to the framework proposed by Németh and Schmidt. The first limitation concerns the logic and structure of the framework. The categories have been formulated as opposites. The answers to variables generate a yes or no answer with the help of specific indicators. The authors use a negative (most controlled) and a positive scale (least controlled). However, many stated effects of the variables could be questioned (see variables 5 and 7, but also 3, 8, and 20 in Table 4.1). This is very apparent in the categories of surveillance–policing (13–17). There is room for debate about whether security measures control or encourage use. In which ways does the presence of art lead to greater control of users? Furthermore, universal access (9–12) is a very commonly identified aspect needed for a functioning public space. Largely stemming from the discourse on the public sphere and its link to democracy and participation, restrictions to people and use are seen as a threat to the public sphere and ultimately sociability and democracy. Interestingly, universal access has hardly been a defining feature of public space through time. On the contrary, public spaces have always been used by closely defined publics (Webster 2002, 397f).[4] It has even been argued that exclusion is a prerequisite for certain kinds of publics (such as minority groups) to manifest (Iveson 2003, 224). Consequently, the effects of controlling and/or encouraging use are neither good nor bad by nature but ultimately depend on the specific context.

The second concern is about its limited applicability, as the framework has been explicitly developed to compare publicly accessible places (either privately or publicly owned) in the United States. The location in a business improvement district or near a commercial building (16, 17) has little relevance for

Table 4.1 Framework developed by Németh and Schmidt (2011) (adapted)

#	*Category*	*Effect*	*Variable*
1	Design – Image	Control use	Design to imply appropriate use
2	Design – Image	Control use	Presence of sponsor/advertisement
3	Design – Image	Encourage use	Restroom available
4	Design – Image	Encourage use	Diversity of seating types
5	Design – Image	Encourage use	Various microclimates
6	Design – Image	Encourage use	Lighting to encourage nighttime use
7	Design – Image	Encourage use	Small-scale food consumption
8	Design – Image	Encourage use	Art/cultural/visual enhancement
9	Access – Territoriality	Control use	Constrained hours of operation
10	Access – Territoriality	Control use	Areas of restricted/conditional use
11	Access – Territoriality	Encourage use	Entrance accessibility
12	Access – Territoriality	Encourage use	Orientation accessibility
13	Surveillance – Policing	Control use	Secondary security personnel
14	Surveillance – Policing	Control use	Security personnel
15	Surveillance – Policing	Control use	Security cameras
16	Surveillance – Policing	Control use	In business improvement district
17	Surveillance – Policing	Encourage use	At a commercial building
18	Law – Rules	Control use	Subjective/judgement rules posted
19	Law – Rules	Control use	Visible sets of rules posted
20	Law – Rules	Encourage use	Sign announcing public space

other geographic contexts. As a consequence, the list of variables reflects this context and would require certain adjustments in order to be applicable to a Brazilian case study. Thirdly, and partly linked to the first shortcoming, the framework focuses on the control effects on use/users and therefore lacks certain variables to describe the environmental control of hybrid public places. Variables would need to be added to capture the artificial nature of air, light, and temperature.

For the above reasons, I would propose certain modifications to the framework of Németh and Schmidt. As a first step, the problematic effect of encouraging use, which ultimately must be established empirically, is eliminated and effect parameters are conceptualized as binaries within a single variable. A variable can have an effect of greater control (C) or openness (O). The effects are also more narrowly understood as changes within the respective category (either design–image, access–territoriality, surveillance–protection, use–activity). As several variables cannot be directly attributed to either effect, I propose to add the effect of enhancement. Such an effect aims at improving the respective category without having a direct consequence on how open or controlled a variable is (such as the presence of arts, or recreational facilities).

In total, the taxonomy consists of four categories of ten variables each (see Table 4.2). Eight variables have been added to capture the spatial reality of shopping malls and other fully enclosed public places (artificial light, environmental and spatial protection, etc.). Another seven are proposed to better account for different uses and five others have been designed to enable the analysis of public space in its larger context, such as its connectivity with a public space network, different transportation modalities, and its physical boundaries (in line of the accessibility parameters of PPS).

Through the assessment of each variable, each category can be established as predominately controlled (C) or open (O). In total 16 different combinations are possible and reproduced in Table 4.3. Shopping malls are a good example of a hybrid space that is situated at the upper end of the scale, providing fully controlled design, access, use, and surveillance. The position of traditional public spaces should be further down with fewer restricted categories. More empirical research is needed to locate real spaces within this continuum. As a preliminary step, this analytical framework will be tested in the city of São Paulo, after outlining historical changes in the perception and manifestation of public spaces within the city.

Public space in the Brazilian debate

The issue of public space and its demise has been vividly discussed in Brazil by academics and policy makers. The ongoing debate puts stress on two different components giving rise to two rather disconnected discourses, both of which are concerned with identifying the underlying causes of the demise of public space rather than analyzing specific places.

The first discourse deals with public space from the angle of political participation, democracy, and social movements (Caldeira and Holston 2005). From this standpoint, public space is equated with the political space where opinions can be expressed and democratic goals pursued. From this angle, the privatization of public space is seen as a threat to democracy and free speech (Kohn 2004).

The second discourse focuses on the issue of socio-economic segregation and the impact of gated communities on the urban fabric, public life, and societal cohesion. Both the national and international literature on gated communities position the emergence of this typology within contemporary urban transformations and globalization. The trigger, it seems, similar to historic examples,[5] is the increasing fear of violence (Low 2004, see also de Souza 2009).

Let us have a closer look at the city of São Paulo in order to better understand the path-dependent development and mutual dependencies of the transformation of public space, the proliferation of gated communities, and the spread of urban violence.

The city of São Paulo and its urban transformation

The city of São Paulo is part of a metropolitan area with a total population of 21.1 million – the largest in Brazil and one of the largest in the world (2014: 10th position). The contemporary city is characterized by a complex and specific level of segregation, which is itself a product of its particular historic development (Marques 2014). According to Reis (2004), the urban development of São Paulo followed four phases: the Colonial and Imperial City (1554–1889), the European City (1889–1930), the American city (with a strong modernist agenda, 1930–1986) and the Democratic City (urban renewal and reaffirmation of urban planning since 1986). While specific periods might differ, this development trajectory is largely aligned with that of other Latin American countries (see also Coy and Töpfer 2014, 102).

For most of its urban history, São Paulo has been a small town with small extensions to accommodate growth. The very moderate rate of urban expansion allowed for the layout and careful crafting of the structure of public space. Two major events in the 17th century and 19th century completely altered the face of the city. Gold was discovered in Minas Gerais and São Paulo served as primary gateway for its trade and the city ascended to be the commercial center of the national coffee production. This attracted a lot of international migration adding to the labor influx stemming from the outlawing of slavery in 1888. The changes in the city reinforced a socio-economic division based on its geography: the commercial center situated around *Rua São Bento* and *15 de Novembro*, the low-lying but hillier parts of the city were occupied by the working class and factories and, finally, the flatter high-grounds accommodated the affluent population.

During the period of the European City (1889–1930), São Paulo underwent significant spatial transformations which led to a socio-geographical correlation

Table 4.2 Proposed public space taxonomy (in the variable column, terms with a star are shared with Németh and Schmidt (2007, 2011)).

#		Variable	Effect	Indicator
		Category: Design–Image		
I	1	Uniform design	Controlled	0 = Uniform design language (materials, elements, etc.)
		Diversity/ absence of designs	Open	1 = No clear design language of the place
I	2	Design to imply appropriate use*	Controlled	0 = Elements restrict walking paths
		Unrestricted use	Open	1 = Seating and resting elements enable various ways of appropriation
I	3	Seating type prescribed by design	Controlled	0 = Seatings have elements to control the way of sitting
		Diversity of seating types*	Open	1 = There are several options to sit and rest differently
I	4	Presence of sponsor*	Controlled	0 = At least one sign announcing the sponsor or owner of the place
		Absence of sponsor	Open	1 = No sign to indicate who owns or maintain the place
I	5	Presence of advertisement/ publicity	Controlled	0 = At least one advertisement and sign of publicity
		Absence of advertisement/ publicity	Open	1 = No sign advertisement and publicity
I	6	Maintained environment	Enhanced	The place does not show any major sign of degradation
I	7	Identifiable place character	Enhanced	The place features at least one unique and distinguishable element
I	8	Art/cultural/ visual enhancement*	Enhanced	There is at least one piece of art or visual attractor
I	9	Waterbodies	Enhanced	The place features a water body
I	10	Vegetation and landscaping	Enhanced	The greenspaces are well maintained and designed
		Category: Access – Territoriality		
#		**Variable**	**Effect**	**Indicator**
T	1	Private ownership	Controlled	1 = The place is owned by a private person or company

		Public ownership*	Open	0 = The place is owned by a public body
T	2	Paid access	Controlled	1 = A fee is levied to enter the place
		Free access	Open	0 = The place is accessibly without fee or obligation to consume
T	3	Defined hours of operation*	Controlled	1 = There are specific hours of operations
		Unconstrained hours of operation	Open	0 = The place is open 24/7
T	4	Car-based access	Controlled	1 = The place is easy to reach by car (no major physical barriers)
		Public transport connectivity	Open	0 = The public mass transportation system services the place
T	5	Continuous path of circulation	Controlled	1 = There is only one way of circulating through the place
		Various circulation options	Open	0 = The place offers various different circulation options
T	6	Disconnected from PS network	Controlled	1 = The pedestrian connection to the place or its quality is poor
		Connected to public space network	Open	0 = The place is conveniently reachable by foot
T	7	Pedestrian zone	Enhanced	More than 50% of the place is a pedestrian zone
T	8	Orientation accessibility	Enhanced	There is at least one well visible orientation aid
T	9	Restroom available*	Enhanced	Restrooms are available for users of the place
T	10	Internet Portals or Wifi	Enhanced	Free Wireless is available

Category: Surveillance–Protection

#		Variable	Effect	Indicator
P	1	Controlled entrance*	Controlled	1 = The entrance is controlled through a camera system or guards
		Uncontrolled entrance	Open	0 = There are no visible signs of entrance control
P	2	Areas of restricted/conditional use*	Controlled	1 = Reserved areas for costumers or otherwise singled out groups
		Unconditional use	Open	0 = No area is restricted in terms of users

P	3	Private security personnel*	Controlled	1 = There is at least one easily identifiable private security personal
		No private security personnel	Open	0 = No sign of private security personal
P	4	Security cameras*	Controlled	1 = At least one camera is visible from the public accessible place
		No Security cameras	Open	0 = There are no identifiable cameras
P	5	Presence of police	Controlled	1 = Police were present at the time of visit
		No police presence	Open	0 = No police were present at the time of visit
P	6	Protective devices	Controlled	1 = Electric wires, barbwire, or pikes are visible from the place
		Absence of protective devices	Open	0 = No electric wires, barbwire, or pikes are visible from the place
P	7	Presence of physical borders	Controlled	1 = Perimeters are made visible through walls, fences, hedges
		Absence of physical borders	Open	0 = There is no visible physical perimeter
P	8	Weather protection*	Controlled	1 = The place is partly (>25% surface) protected from sun and rain
		Open to the sky	Open	0 = The place has almost no (<25% surface) weather protection
P	9	Artificial day lighting*	Controlled	0 = The lighting of the place is largely artificial
		Natural day lighting	Open	1 = The place receives largely natural light
P	10	Controlled climate	Controlled	1 = The place has a temperature controlled climate
		Natural climate	Open	0 = The climate of the place is subject to natural calamities

Category: Use–Activity

#		Variable	Effect	Indicator
U	1	Behavioral rules posted*	Controlled	1 = Rules about allowed and forbidden activities are made visible
		No behavioral rules posted	Open	0 = No behavioural rules could be found
U	2	Defined activities	Controlled	1 = The place does not explicitly allow informal activities

		Alternative activities allowed	Open	0 = Informal activities could be observed during site visit
U	3	No political expression allowed	Controlled	1 = Political expressions are explicitly forbidden
		Political expression allowed	Open	0 = Political expressions are not explicitly forbidden
U	4	Restauration: Food*	Enhanced	There are places for food production and consumption
U	5	Restauration: Drinks	Enhanced	There are places for drinks and snacks
U	6	Commercial use: Convenience	Enhanced	There are shops selling convenience goods
U	7	Commercial use: Non-Convenience	Enhanced	There are shops selling non-convenience goods
U	8	Non-commercial use allowed	Enhanced	The place featured arts or other performances
U	9	Recreational activities	Enhanced	There is an infrastructure to enable recreational use
U	10	Playgrounds and children animations	Enhanced	The place offers playground and children animation

Table 4.3 The taxonomy of public spaces and its hybrid forms

Category	*1*	*2*	*3*	*4*	*5*	*6*	*7*	*8*	*9*	*10*	*11*	*12*	*13*	*14*	*15*	*16*
Design–Image	C	O	C	C	C	O	C	C	O	C	C	O	O	O	C	O
Access–Territoriality	C	C	O	C	C	O	C	O	C	O	C	O	O	C	O	O
Surveillance–Protection	C	C	C	O	C	C	O	O	C	C	O	O	C	O	O	O
Use–Activity	C	C	C	C	O	C	C	C	O	O	O	C	O	O	O	O

of different population groups in concentric rings around the center. The wealth generated by the coffee boom was invested to make São Paulo the new industrial center of the country. While these works completely changed the face of the city, they foremost benefited the modernization of the center and urban elites. The other parts of the city were characterized by poor environmental conditions and a lack of public services. In the 1920s, the Plan of Avenues created large street corridors linking the center to the outskirts. The residences of the affluent population were established along these infrastructural axes. *Avenida Paulista* became famous for the palatial houses of the local coffee barons, while *Rua Augusta* emerged as an upper-end commercial area (Villaca 1998). This spatial logic remained largely intact during the following two decades.

During the period of the American city, and particularly between 1950 and 1970, we witness a spatial explosion of the city triggered by the economic miracle of Brazil and the heightened migration flows from other parts of the country. The pattern of urbanization continues to follow the classic center–periphery model with affluence in the center and along a South-West axis (*Bela Vista, Jardims*), with poverty concentrated in the periphery. The demographic growth has largely taken place in the poor areas, which added twice as many new residents in this period than the city core.[6] Urban planning remained concentrated on the requirements and spaces of the urban elites. The historic center faced increasing difficulties to accommodate the demands of the booming service sector. A new business center, the *Avenida Paulista,* was created south-westward of the crowded city center. This growth direction also explains later shifts in centrality during the 1990s and 2000s towards *Faria Lima, Itaim* and *Marginal Pinheiros*.

Poverty and deprivation were traditionally concentrated in outlying areas (Bonduki and Rolnik 1982), which partially started to change in the course of the 1960s when real estate market practices and the development of main infrastructure lines opened up peripheral land to the interests of urban elites and the growing middle class. Developers promised full-serviced urban neighborhoods aligned to their lifestyle aspirations. A good example is *Alphaville* in the north-western area of the city which has been developed since the 1970s and is now a current day edge city with 40,000 inhabitants (see Garreau 1991).

Like American development, this process of residential segregation was accommodated by the emergence of shopping centers which provided new centralities for the new suburban residents. São Paulo constructed one of the first shopping centers in Latin America (1966), its spatial expansion multiplied during the 1980s and 1990s. Following its North-American model, these shopping centers included an increasing array of recreational offers, including ice skating rinks and roller coasters. The success of peripheral residential communities and enclosed shopping malls is paralleled by the degradation of the city center. The largely abandoned traditional public spaces were occupied by informal economic activities and poorer income population. Middle and higher income groups perceived these spaces as unorganized, deteriorated, and even dangerous (see Frúgoli Jr 1995, 37ff).

The development of the Democratic City revealed major urban problems. The pull of shopping centers to abandon traditional public spaces was reinforced by a push of spreading urban violence since the early 1980s (de Morais 1985). The lost decade and economic stagnation of the 1990s led to the proliferation of favelas, drugs, and violent conflicts. In the late 1990s, São Paulo was among the deadliest cities in the world with a homicide rate of 37.7 per 100,000 inhabitants. As a consequence, inhabitants stopped visiting certain spaces, avoided others at certain times, and accepted the installation of protective measures in the spaces they frequented (de Souza 2014, 153f). The demise of traditional public spaces was accelerated, favoring the attraction of new hybrid forms, characterized by a high presence of security infrastructure, including fences, walls, gates, surveillance cameras, and guards.

Despite a considerable improvement in the security level of the city, the process of gating continues unabated, pointing at a more complex relationship. The actual link between violence and gating is relatively weak as evidenced by empirical research around the world (Glasze, Webster, and Frantz 2006). It is the feeling of insecurity, as subjective as it may be, that pushes people to seek the refuge of enclosed urban environments. But an equally important explanatory vector in Latin America (the pull factor) is the desire to access reliable infrastructure services and urban environmental quality (Villaca 1998). The urban elites and the growing middle class, with aspirations oriented towards the Global North, have elevated demands for these basic civic services and see private condominiums as a solution to satisfy them.

Nowadays, São Paulo is a complex configuration of a broadly diversified offer in terms of quality and geographic scales. From *fevalas* to luxury mansions, from high-end full-service malls to degraded and dangerous public spaces which even the police have stopped entering, the neoliberal logic has inscribed itself into space, and decides not only where you live but also where you work and shop. Gated communities have been created to cater for the security demands of a growing middle class, and are increasingly spreading to lower income neighborhoods as well. Consequently, gating becomes inscribed into socio-economic reality and ultimately into space. Segregated cities offer parallel life worlds with few contact points between the rich, middle class and the poor. The sustainability of such practices can be questioned as broadly segregated cities may offer few possibilities for developing a public sphere. This established socio-economic order of parallel lifestyles is more psychological than as a result of access control, but its absoluteness is well illustrated by the fierceness of public debates, when this order is called into question (such as organized shopping mall "invasions" by the youth living in the poor periphery).

To summarize, we can state that the Brazilian experience largely follows the pattern of American spatial transformation, although (the fear of) violence appears to have played a greater role. As a consequence, gating has spread broader in urban space and penetrated deeper into Brazilian society. Nowadays we have gated communities on different scales, within several urban functions, and across different socio-economic layers spanning from the urban district to the individual house and from the affluent to the urban poor. The advanced segregation of the city has made parallel life worlds possible where movements of people follow spaces designed for their specific socio-economic reality. Consequently, public space has not only been hybridized but also fragmented. Recalling the discourse on the public sphere, society requires places where difference meets (seen, being seen) and interacts in order to develop civility and democracy. Such encounters are increasingly difficult in the city of walls. If public spaces are indeed mirrors of society, what does this say about our level of civility and democracy?

Testing the taxonomy: a comparison of two public spaces

In the final section of this chapter, I will test the taxonomy of public spaces proposed in the first part of the chapter. The framework of four categories each with ten variables can be applied separately to evaluate the design/image, access/territoriality, surveillance/policing, or law/rules of a given place. In combination, the categories help to assess the control level of a publicly accessible space. Two case studies in the city of São Paulo have been selected for the test analysis. The shopping mall, *Vila Olympia*, and the square, *Praça Silvio Romerio* were chosen for the representative nature of their given typologies. In other words, and keeping in mind the aim of assessing different levels of publicness, there is nothing special about these two cases.

Both places were visited in July 2016 and January 2018. Data were collected through observations and semi-structured interviews with users and business owners in direct vicinity of the publicly accessible spaces. This workflow shares similarities to the research design of other public space models that are based on relative and quantified scales (e.g., Varna and Tiesdell 2010, 592f; Németh 2009, 2485). This considerably simplifies the collection of data as the majority of variables can be directly accessed through observations. Where necessary site plans were consulted and/or place users and employees of nearby businesses interviewed. With this collected data it is possible to calculate the numeric score of closeness.

The publicly owned (T1)[7] square is located in *Tautuapé*, not far from the metro line station (800 m) and two major shopping malls. In addition, several bus lines stop at the square (T4-O). While the public transportation is well developed, the square is rather disconnected from pedestrian networks (T6). The road leading to the metro station is sloped and degraded. Also, the square itself shows considerable signs of degradation and is badly maintained (I-6). While the public space is surrounded by commercial stores and restaurants (U 4–7), the stores finishings are mostly composed of cheap materials. Security

Category	1	2	3	4	5	6	7	8	9	10	Σ
Design - Image	O	O	O	O	C	-	E	-	-	-	O
Access - Territoriality	O	O	O	C	O	C	E	-	-	-	O
Surveillance - Protection	O	O	O	C	C	C	O	O	O	O	O
Use - Activity	O	O	C	E	E	E	E	-	E	-	O

Category	1	2	3	4	5	6	7	8	9	10	Σ
Design - Image	C	C	C	C	C	E	E	E	E	-	C
Access - Territoriality	C	O	C	C	C	C	E	E	E	E	C
Surveillance - Protection	C	C	C	C	O	O	C	C	C	C	C
Use - Activity	C	C	C	E	E	E	E	E	E	E	C

Figure 4.1 Comparison of two different public spaces

infrastructure in the form of cameras (P4), spikes, and barbed wires, are omnipresent (P6). At the time of visit, police forces were stationed there (P5). The public square features controls on all variables that visually emphasize surveillance. On all other categories, particularly on design and use, the common effect is openness. Overall, the square scores very high on openness (17 of 24) and low on control (7 total) as well as enhancement (7 of 16).

The shopping mall is situated in the neighborhood of the same name. It has a comparable distance to the closest node of public transportation (urban train line) as the former example but is clearly conceptualized for car-based access (T4-C). The location of this upmarket shopping mall is very strategic: it is in very close proximity to *Itaim Bibi*, one of the new business districts of the city but also to the posh neighborhoods of *Moema* and *Jardins*. The mall is a multi-floor, fully enclosed and environmentally controlled shopping mall (P7–10) with specific opening hours (T3). It offers upscale stores, restaurants (U4–6) and entertainment (including a movie theater; U9). Children play areas and animation programs are offered to attract young families.

The taste of the target audience is also reflected in the design (spectacular escalator composition; I7) and choice of lavish materials. There is a clear distinguishable design language (I1) with clearly allocated seating options (I3). The environment is kept clean and shows no sign of degradation (I6). An information point and signs enable easy navigation (T8). Private security personnel (P3) and surveillance cameras (P4) are present on all floors but kept out of direct sight. In combination with strict access control (P1) but a general lack of other signs of protective devices (P6-O) the place may appeal to its user due to its high security level without leaving much visual trace. The mall is a highly restricted environment (21 of 24 variables) offering a full range of enhancements.

The two examples show two extreme cases of control in publicly accessible spaces. Although the two examples differ in the level of control and openness, neither is fully controlled or opened. The 40 variables allow for the identification of differences but also commonalities (such as various variables of surveillance and protection) between the two spaces. Such insights connect well to the historical outline and the discourse of the "city of fear": security measures are well deployed (but not equally visually present) in both public spaces.

While these results are analytically promising, more research is needed to fully validate the usefulness of the proposed taxonomy. More (detailed) case studies could help the taxonomy of public spaces and reveal whether certain controlled-open combinations do not already exist in practice. Once all empirical values are placed, it would be worthwhile trying to evaluate the transferability of certain variables and their impact on the use of space and the quality of the public sphere.

Conclusion

This chapter has discussed the historic development of public space in the city of São Paulo by introducing the general importance of public space and its most important academic debates.

The historic development of São Paulo has shown a strong segregative tendency, with spatial displacements of urban centrality and its public spaces. This tendency has been aggravated since the 1960s through the technological advancements of car-based mobility and shopping malls, and particularly since the 1980s through the proliferation of urban violence and gated communities. The pull and push factors have changed over the last two decades which, in return, did not affect the general trend towards greater socio-spatial separation. Successful public spaces are strange hybrids with private ownership, high-levels of control and surveillance to guarantee the right kind of "costumers" and the exclusion of certain publics – privacy and freedom are traded in for a feeling of safety.

After reviewing recent changes in the academic debate, I developed a new approach based on the framework of Németh and Schmidt but considerably adjusted and extended to better capture the complex reality of hybrid public spaces. The proposed taxonomy has four categories, each category has ten variables in order to account for the role of control on the physical characteristics of a place but also its possible use. Like the argument of Németh and Schmidt, my case study suggests that privately owned and managed spaces are successful due to a high level of control. Deeper analysis of public spaces, their level of privatization, and the structure of the city are needed to understand the popularity of shopping malls and other gated forms and the demise of classic forms of public space.

The required next steps include a considerable expansion of the case studies in the city of São Paulo, to further test the applicability of the proposed framework. This should be done by including a broader range of hybrid spaces such as corporate plazas, parks, and shopping streets. This research is currently under way and aims to collect data for a total of 50 publicly accessible places. This larger case study will allow for further adjustments to the categories and variables presented here.

In parallel, the research on publicly accessible spaces ought to be expanded thematically as well as geographically. Thematically, the most pressing need is a better understanding of users to analyze what kind of different publics are using specific places as well as those spaces' quantitative and qualitative changes during different hours of the day and days of the week. These insights are needed to start drawing connections between the ownership and management of publicly accessible places, their level of control, and their impacts on users.

Geographically, it would be interesting to expand the public space investigation in a first step to other Brazilian cities and in a second step to other parts of Latin America. This could help to identify place specificity and better assess the contribution of historical developments in the way public space manifested in a city. This could answer the question whether public space hybridization follows a specific Latin American trend or if there are distinguishable differences between countries.

Acknowledgements

The research idea was conceptualized and developed during my visiting fellowship at the research area The City as Social Space of the Research Institute for Regional and Urban Development (ILS) in Dortmund. I would like to thank all my colleagues for their time and input, and in particular Ralf Zimmer-Hegmann for granting me this opportunity. Furthermore, I would like to extend my gratitude to the anonymous reviewers and the book editors for their thoughtful comments and suggestions. These helped to considerably improve the quality, structure, and argument of the chapter. All remaining errors are my own.

Notes

1 Jane Jacobs attacked fiercely the urban planning ideology of her time that advocated garden city ideas not only to plan suburban developments but also the city itself. According to her, low-density developments have detrimental effects on public life, urban economies, and safety (Jacobs 1961).
2 Urban developments in Latin America have produced highly gated housing solutions (such as the colonial patios), neighborhood, and even town solutions (such as worker towns for mining activities, Janoschka and Borsdorf 2006, 97).
3 There are two major discourses, one focusing on the political dimension (public sphere), the other stressing the spatial dimension (Low and Smith 2006, 6). Critical essays on the *spaces* of the public and its demise started in the United States in the 1960s. Books by Kevin Lynch and Jane Jacobs on the evolution of public space and its symbolic meaning and the role of modern urban planning paradigms were written in the context of ongoing sub-urbanization of American cities (Lynch 1960; Jacobs 1961).

4 The often-cited Greek agora only functioned by limiting its social and political community to free men, excluding slaves and women (Mitchell 1995, 116). A similar logic of exclusion can be retraced for ancient Rome, and Medieval towns or public spaces in cities prior the industrial revolution. Settlements were always places with different types of people, and community limits were also established.
5 The link between urban violence and the consequential avoidance of public space has been described for ancient Rome in the Antiquity and 18th century London (Tuan 1979).
6 http://earthobservatory.nasa.gov/IOTD/view.php?id=83987.
7 I indicate in parentheses the position this variable can be found in Table 4.2. The first letter shows the category, e.g. T for access/territory while the number helps to locate the respective variable, in this case T1 is private ownership.

References

Banerjee, Tridib. 2001. "The Future of Public Space: Beyond Invented Streets and Reinvented Places." *Journal of the American Planning Association* 67(1): 9–24. doi:10.1080/01944360108976352.

Blakely, Edward J., and Mary Gail Snyder. 1997. *Fortress America: Gated Communities in the United States*. Washington, DC: Brookings Institution Press.

Bonduki, Nabil, and Raquel Rolnik. 1982. "Periferia Da Grande São Paulo: Reprodução Do Espaço Como Expediente de Reprodução Da Força de Trabalho." In *A Produção Capitalista Da Casa (E Da Cidade) Do Brasil Industrial*, edited by Ermínia Maricato. São Paulo: Alfa Omega Ltda.

Caldeira, Teresa, and James Holston. 2005. "State and Urban Space in Brazil: From Modernist Planning to Democratic Interventions." In *Global Assemblages: Technology, Politics, and Ethics as Anthropological Problems*, edited by Aihwa Ong and Stephen J. Collier, 393–416. Malden, MA: Blackwell Publishing.

Carmona, Matthew. 2010. "Contemporary Public Space, Part One: Classification." *Journal of Urban Design* 15(2): 265–281.

Carmona, Matthew. 2015. "Re-Theorising Contemporary Public Space: A New Narrative and a New Normative." *Journal of Urbanism: International Research on Placemaking and Urban Sustainability* 8(4): 373–405. doi:10.1080/17549175.2014.909518.

Cohen, Lizabeth. 1996. "From Town Center to Shopping Center: The Reconfiguration of Community Marketplaces in Postwar America." *American Historical Review* 101 (4): 1050–1081. doi:10.2307/2169634.

Coy, Martin, and Tobias Töpfer. 2014. "Inner-City Development in Megacities between Degradation and Renewal: The Case of São Paulo." In *Megacities: Our Global Urban Future*, edited by Frauke Kraas, Surinder Aggarwal, Martin Coy, and Günter Mertins, 101–120. Dordrecht: Springer.

De Morais, Regis. 1985. "O Que É Violência Urbana." *Coleção Primeiros Passos. Brasiliense* 42.

De Souza, Marcelo Lopes. 2014. "Phobopolis: Violence, Fear and Sociopolitical Fragmentation of the Space in Rio de Janeiro, Brazil." In *Megacities: Our Global Urban Future*, edited by Frauke Kraas, Surinder Aggarwal, Martin Coy, and Günter Mertins, 151–164. Dordrecht: Springer. doi:10.1007/978-90-481-3417-5_11.

De Souza, Marcelo Lopes. 2009. "Social Movements in the Face of Criminal Power." *City* 13(1): 26–52. doi:10.1080/13604810902770788.

Frantz, Klaus. 2006. "Private Gated Neighbourhoods: A Progressive Trend in US Urban Development." In *Private Cities: Global and Local Perspectives*, edited by Georg Glasze, Chris Webster, and Klaus Frantz, 61–72. New York: Routledge.

Fuchs, Anthony, Igor Guatelli, Jaap Klaarenbeek, and Jasper Moelker. 2009. "The Open City – A Matter of Access?" Paper presented at the N-Aerus Conference, Rotterdam, Netherlands, October 1–3. http://n-aerus.net/web/sat/workshops/2009/Rotterdam/pdf/Fuchs_etal.pdf.

Garreau, Joel. 1991. *Edge City: Life on the New Frontier*. New York: Doubleday.

Glasze, Georg, Chris Webster, and Klaus Frantz. 2006. *Private Cities: Global and Local Perspectives*. London, New York: Routledge. doi:10.1007/s13398-014-0173-7.2.

Harteveld, Maurice G.A.D., and Anthony Fuchs. 2007. "A Forgotten Chapter in European Urbanism: The Invention of the Mall, Its Mutation and Inherit Potentials." Paper presented at the 4th International Seminar on Urbanism and Urbanization, TU Delft, September 24–26.

Harteveld, Maurice G.A.D. 2006. "Viva Las Vegas: A Search for the Urban Design Task of Interior Public Space." In *Urban Transformations and Sustainability: Progress of Research Issues in Urbanism 2005*, edited by Frank van der Hoven and H.J. Rosemann, 32–41. Delft: Delft University Press.

Iveson, Kurt. 2003. "Justifying Exclusion: The Politics of Public Space and the Dispute over Access to McIvers Ladies' Baths, Sydney." *Gender, Place & Culture* 10(3): 215–228.

Jacobs, Jane. 1961. *The Death and Life of Great American Cities*. New York: Random House.

Janoschka, Michael, and Axel Borsdorf. 2006. "Condominios Fechados and Barrios Privados: The Rise of Private Residential Neighbourhoods in Latin America." In *Private Cities: Global and Local Perspectives*, edited by Georg Glasze, Chris Webster, and Klaus Frantz, 89–104. Abingdon: Routledge.

Kohn, Margaret. 2004. *Brave New Neighborhoods: The Privatization of Public Space*. New York: Routledge.

Lopes, Miguel Nogueir, Sara Cruz, and Paulo Pinho. 2012. "The Changing Publicness of Urban Spaces." Paper presented at the European Symposium on Research in Architecture and Urban Design (EURAU 12), Porto, Portugal, September 12–15.

Low, Setha, and Neil Smith. 2006. "Introduction." In *The Politics of Public Space*, edited by Setha Low and Neil Smith. New York: Routledge.

Low, Setha. 2004. *Behind the Gates: Life, Security, and the Pursuit of Happiness in Fortress America*. New York: Routledge.

Lynch, Kevin. 1960. *The Image of the City*. Boston, MA: MIT Press.

Madanipour, Ali. 2003. *Public and Private Spaces of the City*. Abingdon, Oxon: Routledge.

Madden, David J. 2010. "Revisiting the End of Public Space: Assembling the Public in an Urban Park." *City & Community* 9(2): 187–207. doi:10.1111/j.1540-6040.2010.01321.x.

Marcuse, Peter. 2005. "The 'Threat of Terrorism' and the Right to the City." *Fordham Urban Law Journal* 23: 767–785.

Marques, Eduardo. 2014. "Estrutura Social E Segregação Em São Paulo: Transformações Na Década de 2000." *Dados – Revista de Ciências Sociais* 57(3): 675–710. doi:10.1590/00115258201421.

Mitchell, Don. 1995. "The End of Public Space? People's Park, Definitions of the Public, and Democracy." *Annals of the Association of American Geographers* 85(1): 108–133. doi:10.1111/j.1467-8306.1995.tb01797.x.

Morris, Anthony Edwin James. 1972. *History of Urban Form: Prehistory to the Renaissance*. London: George Godwin.

Németh, Jeremy. 2009. "Defining a Public: The Management of Privately Owned Public Space." *Urban Studies* 46(11): 2463–2490. doi:10.1177/0042098009342903.

Németh, Jeremy, and Stephan Schmidt 2007. "Toward a Methodology for Measuring the Security of Publicly Accessible Spaces." *Journal of the American Planning Association* 73(3): 283–297.

Németh, Jeremy, and Stephan Schmidt. 2011. "The Privatization of Public Space: Modeling and Measuring Publicness." *Environment and Planning B: Planning and Design* 38(1): 5–23.

Newman, Oscar. 1972. *Defensible Space*. New York: Macmillan.

Newman, Oscar. 1996. *Creating Defensible Space*. US Department of Housing and Urban Development, Office of Policy Development and Research. Institute for Community Design Analysis, Center for Urban Policy Research, Rutgers University, Washington, DC.

PPS (Project for Public Spaces). 2000. *How to Turn a Place Around: A Handbook for Creating Successful Public Spaces*. New York: Project for Public Spaces.

Reis, Nestor Coulart. 2004. *São Paulo: vila cidade metrópole*. BankBoston.

Sennett, Richard. 1992. *The Fall of Public Man*. New York: W. W. Norton & Company, Inc.

Sorkin, Michael. 1992. *Variations on a Theme Park: The New American City and the End of Public Space*. New York: Macmillan.

Tuan, Yi-fu. 1979. *Landscapes of Fear*. Minneapolis: University Minnesota Press.

Varna, George, and Steve Tiesdell. 2010. "Assessing the Publicness of Public Space: The Star Model of Publicness." *Journal of Urban Design* 15(4): 575–598. doi:10.1080/13574809.2010.502350.

Villaca, Flavio. 1998. *Espaco Intra-Urbano No Brasil*. São Paulo: Studio Nobel.

Webster, Chris. 2002. "Property Rights and the Public Realm: Gates, Green Belts, and Gemeinschaft." *Environment and Planning B: Planning and Design* 29(3): 397–412.

Weintraub, Jeff. 1997. "The Theory and Politics of the Public/Private Distinction." In *Public and Private in Thought and Practice: Perspectives on a Grand Dichotomy*, edited by Jeff Weintraub and Krishan Kumar, 1–42. Chicago, IL: University of Chicago Press.

Young, Iris Marion. 1990. *Justice and the Politics of Difference*. Princeton, NJ: Princeton University Press.

Chapter 5

A window into the soul

Biosensing in public

Elaine Sedenberg, Richmond Wong and John Chuang

Introduction

There is a growing public consciousness of the information trails and digital traces generated throughout our everyday lives – keystrokes, smartphones, electronic records, and interactions with digital interfaces leave millions of breadcrumbs outlining our physical paths and generate opportunities for social inferences (see, e.g., Madden and Rainie 2015). But what about our physiological and emotional states? Our heart rates, body temperatures, expressions, tones, involuntary reflexes, and biological signatures make us human yet expose our innermost nuanced physical and mental states. Though these signals have existed throughout human time, this potential for sensors to continuously record *en masse* and to systematically analyze our innate biological nature uncovers a new sensitive data class: biosensed information.

Many scholars have begun grappling with privacy and the ethical implications stemming from the quantified-self movement and the wearable industry's collection of these data (see, e.g., Crawford, Lingel, Karppi 2015). Though biosensing is commonly associated with wearables or sensors embedded into "smart" objects, little attention has been paid to the potential of these biosignals to be captured publicly from a group of non-consenting individuals over extended timescales – either remotely, through casual contact, or from biological traces left behind. Consumers of wearables have agency to put on or take off a wearable they own, while biosensing in public – outside of the consumer framework, thus bypassing consent, at least in part – frustrates our current privacy norms. Even though many of the same issues – integrity, accidental disclosure, inferential potential, generation of new knowledge, etc. – are shared across both device owners and general data subjects, the lack of consent, leaky information flows, and potential for large-scale collection and analysis by a wide spectrum of actors makes this a particularly thorny issue.

This chapter examines the unique nature and inferential potential of biosensed data more broadly, to understand the urgency to consider how these data may be collected and used to create new privacy concerns, particularly in public. It then maps a subset of current biosensing technologies that may collect data remotely, via casual contact or from traces left behind, to outline the current capabilities of biosensing technologies and root potential privacy harms within current technical capabilities. Since static and video images may be used to deduce many biosignals remotely, this chapter analyzes the past evolution of social norms to mitigate or accept new privacy threats posed by developments in camera technology. Understanding the role of past social norms in mitigating harms posed by portable, consumer-grade camera devices helps illuminate the potential role norms may play within a biosensing context. Specifically given this shift in information potential without a shift in underlying technology, this chapter uses Nissenbaum's (2010) "contextual integrity" privacy heuristic to fully explore the need for new norms and remedies for biosensing in a commercial public context. This analysis exposes the need to reassess the conceptual privacy risk models beyond informational privacy. Based on this examination of social norms, remedies, and exploration of biosensing's privacy risk models, this chapter examines relevant legal frameworks in the European Union (EU) and United States (US) to establish the extensibility and potential gaps when applied to remote biosensing privacy harms. Finally, this chapter concludes with a brief outline of possible legal or social remedies that may address privacy needs in public with biosensing technologies.

Certainly, many of the technologies and capabilities are just now on the cusp of development (past laboratory prototypes or proof-of-concept analytics): this chapter calls attention to and anticipates privacy concerns relating to these emerging technologies and systems, examining what is not quite yet reality, but is far beyond mere speculation.

The nature of biosensed data

Biosignals represent an emerging class of data with the potential for massive, systematic, and remote or casual collection of intimate information. The proliferation of sensors makes detection, interpretation, and inference of these previously subtle – or otherwise invisible – emotional and physiological signals possible from proximate and remote locations. Although these cues are essential to biological survival and have existed for millennia, technology stands to sense and collect these data on orders of magnitude more perceptive than the human eye, and on extended timescales impossible through casual observation.

Biosensing technologies are being developed to capture a wide range of physiological signals from which inferences may be drawn, including our mental and emotional states, predispositions to diseases, or proclivities to particular behaviors. Many of these states are unknown or subconscious to us, and many of these patterns are not yet fully understood by modern science. The applications and business models emerging from biosensed information will raise new security and privacy challenges that are not yet fully comprehended or anticipated.

This new emerging class of data sets itself apart from existing information types through a set of unique characteristics.

- **Expansive in scope**. Measurable signals are diverse and include physiological (e.g., electroencephalogram (EEG), electrocardiogram (ECG), heart rate (HR), electromyography (EMG), electrodermal activity (EDA, otherwise known as Galvanic Skin Response or GSR), functional near-infrared spectroscopy (fNIRS), blood pressure, blood glucose) and kinesthetic (e.g., accelerometry, eye-tracking, facial expression recognition). From combinations of these signals, inferences about the body or actions can be drawn to deduce emotional states, physical reactions, and even memories and thoughts.
- **Intimate yet leakable**. Biosensors may be wearables (e.g., commercial fitness trackers worn on the wrist, which require opt-in and are removable) as well as remote (e.g., heart rate monitoring through surveillance video feeds, which do not require contact or consent and which may be invisible to the data subject), and they may be standalone or embedded in everyday objects. Given biosignals may be naturally – and largely uncontrollably – emitted as signatures (e.g., electrical, chemical, or biological) or as involuntary impulses, the information flows are leaky in the sense that diverse actors (or multiple actors at once) may sense them remotely or through brief contact.
- **Precise yet ambiguous**. Biosensors are increasingly able to produce readings with many significant figures, yet the high-level inferences drawn from these raw signals will be context dependent and highly ambiguous. For instance, a single emotion may have different associations or spectrums for one individual compared to others.
- **Familiar yet unverifiable**. When shown a photo, audio clip, or geolocation tag of themselves, it is usually fairly straightforward to confirm these data match the ground truth of a data subject's experience. However, if shown a biosignal, such as an EEG recording, it would be much more difficult to confirm its provenance relating to one's self, not to mention verify its accuracy. Further, experiences may not always map to the measurements: one may feel a racing heart at the sight of a romantic crush, but it may or may not correspond to an increase in heart rate.
- **Limited controllability**. Many of these signals, like those associated with respiration, or involuntary reflexes are controlled by the autonomic nervous system and are difficult to consciously control. Our biological signals, like fingerprints or genetic profiles, are similarly difficult to control since individuals leave traces through casual contact. Some of these signals, impulses, and signatures may be obscured through conscious overriding (e.g., deep breathing) or obfuscated by physical barriers.

Unlike many other information flows that require computer-mediated interaction, biosensed data are emitted as (or may be inferred from changes in) thermal and electromagnetic radiation, visual or observable cues, biochemical signals and traces, or encoded in biological signatures. While sensors may be voluntarily worn by consumers or self-recorded and collated by quantified self-enthusiasts, developing and emerging technologies enable even more data streams to be collected remotely, through casual contact – meaning sensors that may be briefly, and perhaps unknowingly, touched through everyday interaction – or through physical traces left behind. This potential makes biosensed data collection with no opportunity for consent a new privacy challenge for individuals in public spaces.

For each potential type of biosensed information, sensors or diagnostics can be operated from a distance to the individual data subject, after the subject has left a trace, or through casual contact that might be mediated by everyday interactions like touching a door handle, or sitting on a bench. This chapter specifically aims to focus on the potential of biosensing technology apart from wearable devices in order to fully grapple with information flows that may involve no prior consent or knowledge from the individual data subject. Remote biosensing may be deployed on top of existing sensor technology – such as cameras – by introducing new analytical processes that uncover previously undetected biosignals, or through the deployment of entirely new sensor technology.

Table 5.1 presents a taxonomy of sensors and potential capabilities by type of signal, measurable biosensed phenomena, inferential potential, and relevance to remote, traceable, or casual contact public collection.

Each measureable phenomena on its own – such as heart rate, temperature, pupil dilation, etc. – may not contain much inferential power since many of these signals will rise and fall in concert to obvious external stimuli. For example, several measures may increase simultaneously if an individual becomes startled. Insights may be inferred more easily, however, when systems change asynchronously. If heart rate speeds up but another measure like blood oxygenation or EDA stays constant, these data may reflect a mental over an autonomic change of state. Additionally, any change in an individual's usual baseline may indicate a notable change. Biometric authentication that conducts facial recognition or fingerprints – which is already being done on some banking apps – would also inherently come with the potential to give private companies access to other biosensed information (Corkery 2016). Given the nature of biosensed data, voluntarily giving access to one feature like an iris scan may enable countless other inferences to be made simultaneously.

The inferential potential, beyond mere physiological responses derived from these sensors and analytics, is only beginning to come to light. Leaning upon small-scale studies – the current state of the art in this area of research – one can understand how these biosensing systems may be able to uncover unknown emotional or physiological states hidden in plain view. Our microexpressions lie underneath outward facades or in fleeting looks, and betray concealed or subconscious emotional states that are too short or subtle to notice with the naked eye (Adler 2015; Li et al. 2015). Outside of biosensing, there are proof-of-concept academic studies that demonstrate an ability to predict health conditions based on language use alone. One research group used language from Twitter (in particular, the use of angry language) to predict heart disease mortality at a community level. Another study conducted a case study analysis of former US Presidents Ronald Reagan and George H.W. Bush to retroactively compare for differences in discourse complexity that may be related to early Alzheimer's disease (Berisha et al. 2015). It is notable that inferences like these are able to retroactively examine existing corpuses to deduce future health states. With biosensing technology, individuals may produce larger, more intimate data repositories that reveal not only more emotional granularity than a public tweet, but will last into the future, making retroactive analyses possible. Young adults may build biosensed dossiers in their twenties that may be relevant in 30 years for unforeseen analytics. For instance, researchers studied the predictive associations of resting heart rate on violent and non-violent criminality using the measurements of over 700,000 Swedish men when they began mandatory military conscription at age 18. Researchers found a low resting heart rate was associated with an increased risk for violent criminality, non-violent criminality, risk of being a victim of an assault, and being involved in accidents as an adult (Latvala et al. 2015). The full potential for insights and the range of possible uses remain unknown, but the intimacy and raw potential are certain.

Sensing and social norms

Often, the introduction of new technologies grates against pre-existing social norms, and societies are given a relatively short timeframe to react and adjust laws accordingly (Tene and Polonetsky 2013). Bicchieri (2000) describes two separate interpretations of social norms: in one norms are descriptive of what is considered normal behavior and, in the other, norms are injunctive of what the public believes should be done for social approval or disproval. Laws may develop to propagate social norms and further prevent social harm – as was the case in seat belt laws – and laws may also develop to create torts or other possible remedies for socially unacceptable harms (Licht 2008).

Biosensed information, particularly remote biosensing to collect data without any physical contact, frustrates many of the norms for recording in public places without individuals' knowledge or consent. This is especially true given that biosensing is entirely new (at least outside of the medical context) and is experiencing rapid technological and analytical development. The following subsections explore two prior cases of remote sensing in public – specifically, the capture of information using photography beginning with the amateur and portable Eastman Kodak Camera, and followed by the Google Glass embedded

Table 5.1 A taxonomy of sensors and potential capabilities

Measurable phenomena or characteristic	*Methods of measurement*	*Example potential inferences*	*Distance and interaction potential*
Body temperature	Infrared thermal radiation	Activity level; health status; acute changes in stress levels	Remote via infrared cameras or infrared thermometers (spot ratio for accurate measurement is several feet away) (Scigiene Corporation, n.d.).
Heart rate	Electromagnetic radiation; radio frequency (electromagnetic fields)	Activity level; health status; changes in excitement or stress	Vital Radio (several feet away, possible through walls) (Adib et al. 2015; Mullan et al. 2015; Wang et al. 2015). Remote and smartphone cameras also may be able to detect pulse and heart abnormalities (Hewitt 2013; Li et al. 2014) (several feet away; eyeshot).
Eye tracking	Infrared and near-infrared non-collimated light reflections	Attention; gaze; drowsiness; traumatic brain injuries	Some eye tracking technology requires a stable head, but new devices do not, or take advantage of other devices on the head (e.g., glasses, headset). Measurement could be taken from a high-resolution camera a few feet in front of the subject.
Retinal scan	Infrared light absorption	Identity; presence of health conditions (e.g., diabetes, stroke, hypertension, heart disease, neurovascular disease, etc.) (MacGillivray et al. 2014).	Close proximity and bright light.
Iris scan/pupil dilation	Camera with infrared illumination or electromagnetic radiation	Identity; arousal; health status (e.g., stroke history, or in children leukocoia, an early sign of retinoblastoma) (Clusmann, Schaller, Schramm 2001; Palca 2014).	High-resolution images of the eye – potentially through proximate but still remote means via other technologies like Virtual Reality (VR) headsets.

Measurable phenomena or characteristic	*Methods of measurement*	*Example potential inferences*	*Distance and interaction potential*
Facial recognition	Electromagnetic radiation	Identity; presence of genetic disorders (Cooper 2014).	Low-resolution images of a face.
Emotion	Electromagnetic radiation	Mental or emotional states.	Video or images of face may be made from a distance to capture microexpressions.
Fingerprints	Image patterns for identifying characteristics	Identity	Traces may be left behind during casual contact, but fingerprints may be sensed through contactless scanning in one second (Biotime Technology 2014).

camera. Historically images could capture expressions and situations, or tie individuals to particular locations – now images (static or video) may unlock new inferences and measurements of various biosignals. Each photography case provides insight into potential social norms that may arise with remote biosensing technologies, and begins to explore how these norms collide and influence laws and policies.

Introduction of the Kodak Camera

The introduction of the Eastman Kodak Camera in 1884 made photography portable and affordable to the general public, and coincided with the rise of sensationalistic journalism and large print circulations (Solove and Schwartz 2015). Photography was brought into homes and family life, and Kodak advertising promoted the technology in the household as a signal of modernity and leisure (Sarvas and Frohlich 2001, 55). However, this new technology opened private spaces and fleeting moments up to permanent scrutiny, in some ways to certify our memories or impressions of others (West 2000, xi). In response to public fears and normative tensions regarding the potential invasion of cameras and relative permanency of images, resort beaches and even the Washington Monument banned their use temporarily until law and social norms caught up (Lindsay 2000).

Published in 1890, the canonical article "The Right to Privacy" by Warren and Brandeis explicitly states: "Thoughts, emotions, and sensations demanded legal recognition," and calls attention to recent inventions and business methods, "instantaneous photographs and newspaper enterprise," as threats to the sacred precincts of private and domestic life (Warren and Brandeis 1890, 195). This article established the need for a new right for privacy, which was not protected under common law but could be modified using torts to redress harms.

The introduction of the camera clearly illustrates a time where social norms were forced to rapidly adapt to the presence of a new ability to record information (images that relayed our facial expressions, locations, social interactions, and actions) about ourselves and others. Similarly, biosensing technology presents a new information to record physiological and emotional data in previously public, invisible, or fleeting spaces – often using the same portable image technology or similarly small, and amateur friendly recording devices. Existing legal frameworks were not prepared to protect and remedy situations brought on by the collision of law with technological advances, and adaptations had to be made in order to keep up with evolving social norms and values around privacy. At present, it seems absurd – if not impossible – to ban photography in public spaces like the Washington Monument. Yet there are still social situations where it would be seen as socially inappropriate to take photos of others, even it is technically legal.

Google Glass

Google Glass, first publicly announced in April 2012 and released to some members of the public in a public beta program in April 2013, provides insight into user and media perspectives on emerging technologies and the privacy concerns people may raise in anticipation of the release of new devices. After the announcement of Glass in 2012, people imagined what it might be like to live in a world with Glass, a world reflected in Glass's portrayal in Google's promotional videos and advertisements and in broader media discourse. Through the process of speculating and anticipating the future with Glass, people began raising concerns about remote sensing and privacy (Wong and Mulligan 2016).

Google's initial conceptual video of Glass was released on April 4, 2012 on YouTube.com. The video is shot entirely in a first person point of view, only showing the viewpoint of a single male Glass user as he makes his way around New York City. It never shows what the actual device looks like, rendering the device invisible. Furthermore, the user is never seen putting Glass on or taking it off, suggesting that the device is worn and turned on all the time, and not taken off or turned off between interactions with it. The video shows the user walking in the street, at a bookstore, and meeting a friend at a food truck. Notably, these are all in public or semi-public spaces. Together, this video begins to portray Glass as a device that seems to fade into the background, is always worn and always turned on throughout the day, capable of recording pictures and video at any time, and used across many contexts by one person.

Against this backdrop, people began discussing concerns about privacy in several ways. First, people expressed concerns about surveillance and surreptitious recording by (other) Glass users, noting that people could be secretly taking pictures of others in varied public places including the subway, the street, airplanes, or at beaches, clubs, and bars (Hill 2012). These concerns were facilitated by the perception of Glass as an invisible, seamless, and mobile technology. Second, people expressed concerns about how data collected by Google through Glass might be used. Some, contemplating Google's business model, suggested that Glass would be used to collect personal data to deliver more advertisements (Thier 2013) or that Glass would share data with other companies like LinkedIn or Facebook to, for example, conduct facial recognition in public (Burns 2013). A third set of concerns was raised about Glass breaching structural social and technical constraints that protect privacy (Surden 2007). Perceiving Glass as mobile, always-on, and blending into the background suggested that prior structural constraints – such as social norms, limited camera battery life, and the visibility of smartphones – that prevented certain types of data collection might change with Glass. Ephemeral public experiences in everyday life might unwillingly and unknowingly become digitally archived. A fourth set of concerns about Glass transgressing social contexts was raised. Glass's mobility and always-on portrayal suggested that it was untethered to context, leading to fears that information could be taken from one context and easily used in another without regard for the social norms that usually govern information flows in a given context.

As time got closer to the launch of the Google Glass beta program, some people began taking private remedies to protect their privacy, including private business owners banning Glass in bars and strip clubs (DiSalvo 2013), and the formation of an online campaign trying to ban Glass called Stop the Cyborgs.

In many ways, biosensing technology is at a similar stage of development – still under development but with growing public exposure. Like with Glass, the people whose biosignals are being monitored and collected may not be able to see the sensors and may not even know when or where this collection is occurring. Biosensing much like Glass also raises questions about data flows – what people, companies, and third parties are able to access the data, and how these data might be used. Biosensing also begins to erode some structural constraints that protected privacy – signals that were previously protected because no technologies or techniques existed to record them can now be recorded. Moreover, networked biosensing technologies have the potential to violate contextual norms. Perhaps, the Glass example also suggests that when new technologies and technology products are announced and publicized, productive public conversations can occur about the social implications of those new technologies.

Remote biosensing in context

After discussing various social norms and responses that are associated with protecting privacy and exploring historical and recent cases with similarities to biosensing technologies, this chapter now turns to examine emerging aspects of remote biosensing that may be problematic. Nissenbaum's contextual integrity heuristic provides a useful framework to compare the changes in data flows afforded by remote biosensing through new analytical power on existing technology and introduction of novel sensors. Contextual integrity holds that rather than framing privacy as control and prevention of information or data flows, privacy is better understood as a set of relations and information flows governed by "context-relative informational norms" (Nissenbaum 2010). That is, different social contexts have varying social norms that govern appropriate flows of information and data – and the violation of these norms suggests the potential violation of privacy. Four key steps emerge as a heuristic from this framework: 1) establish the prevailing social context; 2) establish the key actors involved in the information and data transfer (including data senders, data recipients, and data subjects); 3) indicate what attributes, or data types, are involved; 4) establish the principles of data transmission. If new practices introduce changes to any of these elements, it suggests a potential violation of contextual integrity. This provides a way to understand and assess how *changes* in practices may affect informational norms.

The contextual integrity heuristic is used here to investigate a plausible deployment of remote biosensing technologies at a shopping mall in the US. While the precise context described in this scenario depicts a situation in a near, proximate future, it is grounded in real, developing technologies, testbeds, prototypes, and pilot projects. By exploring these possibilities while technologies and systems are still in development, concerns related to privacy and discussions of potential remedies can be raised in a proactive, rather than reactive manner, at a time when systems may still be amenable to change. This follows traditions across many fields, most notably in recent calls for "privacy by design," which advocates for addressing privacy concerns through the engineering, design, and use of technologies through technical, business, and legal means (Cavoukian 2011; Computing Community Consortium 2015; Mulligan and King 2011; Rubinstein 2011).

Social context

Imagine a number of remote biosensors from Table 5.1 being used at a shopping mall somewhere in the US. This includes infrared cameras that can detect people's body temperatures, cameras and wireless devices that can sense heart rate, and video cameras that can scan faces for pupil dilation, facial recognition, or emotion. In this example, a user- or customer-centric point of view is taken, focusing on practices that include the sensing of customers. Uses of these remote biosensing technologies might include presenting marketing and advertisements to customers, monitoring for security, collecting data for analysis, one-party actions (such as one user sensing a second user's signals without their knowledge or consent; or a customer service representative sensing a customer), or two-party interpersonal actions (such as two users sensing and sharing their signals with each other). While shopping malls are privately owned, shopping malls are still semi-public spaces, outside the sites of the private domestic home or private workplace.[1] Prior research on human–computer interaction and computer mediated social interaction in public space tends to take a broader view of public spaces, including semi-public sites such as cafes, bars, train stations, and restaurants, which serve the same social functions of public space of sociality and recreation (Humphreys 2005; Humphreys 2010). From a user or consumer point of view, these sites constitute being "in public." Thus following contextual integrity's focus on the perspective of users, this example similarly considers a broader view of "public space" to include semi-public sites of social interaction, such as shopping malls.

Actors

The data subjects in this context are the mall customers. Their data is sensed and collected, although they may or may not consent to the collection of their data (or may consent without full knowledge of the capabilities of technical sensors, inferences, or full knowledge of legal protections they may or may not have). The data collectors vary, but include the mall operators, mall security, store workers and, potentially, other customers that have access to these sensors. Data recipients include all the data collectors, but also include the companies that process and return the data collected by sensors, and potential third party services and companies that store, access, or further process the data.

Attributes

In a mall today, without remote biosensors, several attributes – or data types – are already in play. Photos and videos of customers are recorded for surveillance purposes. Customers' real-time locations can be collected and aggregated using cell phone data (Clifford and Hardy 2013). During financial transactions, customers may share personal information such as their names, addresses, and credit card numbers. With remote biosensed information, additional attributes are invoked as well – both those collected directly by the sensors and those that may be inferred from that data. These sensors may also collect photographic or video images, but at resolutions that allow pupil measurement and the inference of one's identity, emotions, arousal, and health status. Body temperature and infrared images can be collected as well, which may be used to analyze activity level and health status. Heartbeat and breathing activity data can also be collected, providing insight into activity level, health status, excitement, and stress.

Principles of transmission

The contextual integrity heuristic asks us to contemplate current baseline norms that provide the principles of transmission and compare those to any changes in the principles of transmission. The principles of transmission are affected by current practices, laws, technological capabilities, and social norms. Drawing on Solove's taxonomy of privacy harms, the processes of "transmission" can encompass multiple stages of the data lifecycle, including data collection, processing, and dissemination (Solove 2006). The analysis of transmission principles in this example draws upon prior research studying user behaviors with technology and attitudes toward data collection and sharing in public (and semi-public) spaces, current practices, current technical capabilities of widely implemented systems, and laws and regulations that signify and help set social norms. While acknowledging that there are a range of behaviors around how individuals perceive and perform their identities and emotions, this analysis focuses on some of the present principles surrounding the stage of collection, then around the stages of data processing and dissemination. It then explores how the collection, processing, and dissemination of biosensed data may change those particular principles.

Currently, surveillance of physical spaces is primarily visual and camera-based (Räty 2010). Furthermore, in research studying users' perceptions of data collection with new augmented reality devices in public, users make comparisons to already existing visual recording technologies, including cellphone cameras and CCTV cameras (Denning, Dehlawi, and Kohno 2014). This suggests that practices and expectations around these existing visual technologies can be used to understand baseline norms that provide principles of transmission. Currently, security cameras can only capture what is within their line of sight, and the cameras' locations are generally visible to customers. Due to the cameras' visibility by virtue of their physical placement as well as, frequently, the presence of posted notices, customers can choose to avoid an area or avoid a camera's line of sight should they not want to be recorded, or customers may elect to change their behavior or self-presentation (Massimi et al. 2010). Video surveillance is also conducted at a resolution that might allow facial recognition – either computationally or by other humans, though with limited accuracy and consistency (Trottier 2014; Introna and Nissenbaum 2009) – but not at a resolution that would allow for iris detection or pupil measurement. These technical limitations of video surveillance create structural protections for privacy, that is, privacy is implicitly protected because certain technical capabilities do not yet exist or impose heavy costs to implement (Surden 2007). These limitations also create normative expectations about what is possible and acceptable to monitor and collect.

In interpersonal interactions inside the mall, people expect that they can remain pseudonymous and not reveal their identity while walking around; in other words, people expect information about their identity to *not* be collected. People expect that they can interact interpersonally without revealing their identity as well, such as asking an employee for help or asking another person for directions. Individuals have control over when to reveal their identity. In these interactions, people's "body language" might be interpreted, but it cannot be measured or independently verified. One's ground truth emotions and mental states are generally considered to be personal and generally vague in nature – even if sometimes outwardly expressed or revealed in fleeting reactions. Prior work has found that people act in accordance to social norms that allow others to have private conversations in public spaces. For instance, Humphreys (2005) finds that often in public spaces, when someone is conducting a conversation on a mobile phone, he or she will turn away or shield their face while other people in their party will occupy themselves with other activities – even if they are able hear the conversation – in order to acknowledge the social norms of private conversation in public. Social norms also limit what types of recording activities are deemed appropriate in public spaces – for instance, nonconsensual, perverse video recordings of others in public spaces are deemed unacceptable socially. In many states in the US, voyeurism laws support these social norms concerning inappropriate uses of cameras (Solove and Schwartz 2015).

Transmission principles regarding the processing and dissemination of data collected in a public space such as a shopping mall include data sharing with third parties only in particular circumstances and maintaining data security when the data are shared. While many people expect to be recorded by CCTV cameras in public spaces (Nguyen, Kobsa, and Hayes 2008), people's expectations suggest that collected data should not be widely shared for processing and dissemination. Users tend to expect that the people with access to recordings of public spaces include security personnel, employees, owners, and the police, but do not expect the material to be distributed to other actors (Massimi et al. 2010). Other studies corroborate this, finding that participants view camera installations in public spaces to be less acceptable if footage is streamed to a remote location (Friedman et al. 2006). While surveillance camera data may be shared with law enforcement (with a warrant), traditional CCTV surveillance camera footage is stored locally.

Today, internet connected (IP-based) security cameras already challenge these transmission principles, as the video data could easily be stored offsite with third parties. These types of distributed architectures also allow for new types of processing, such as automated or crowdsourced facial recognition (Introna and Nissenbaum 2009; Trottier 2014). These emerging forms of data processing may create changes to the transmission principles, by allowing new actors access to the data, violating subjects' reasonable expectations, or changing how privacy risks and benefits are distributed among people and groups (Introna and Nissenbaum 2009).

When interacting with retailers, customers expect their personal financial information – such as credit card number, name, and address – to be processed by external institutions, but this is done in the context of a transaction, such as a purchase. Engaging in this transaction provides implicit consent for the transmission and use of this information for purposes of the transaction. Customers may also expect the store to collect purchase data to understand customers' habits, possibly using a third party to process the information. The results of this information processing may be used as feedback to try to emotionally manipulate customers' purchasing habits through pricing strategies, "sales" signs, physically arranging displays in particular ways, or using music or aromas in the store (Klosowski 2013). However, this data processing (and resulting feedback interventions) are done at aggregate levels, either on all customers or on broad demographics such as age or gender. Data subjects also generally expect that data security precautions are taken during the processing of their data. Principles of data security are also established through technical frameworks and standards that companies follow, such as the ISO 27001 standard (ISO 2017) or NIST Cybersecurity Framework (NIST 2014).

The introduction of remote biosensing technologies into the mall setting may disrupt these principles and norms. Biosignals are *intimate yet leakable*, allowing remote biosensors to collect the data without sustained contact and without a line of vision, such as Vital Radio, which wirelessly detects heartbeats. Sensors that are not cameras may not look like sensors either. Without these visual cues, customers may not be aware that the sensors even exist or that their signals are being collected, processed, and disseminated – by the mall, by stores and their employees, or by other individual customers. The uses of biosensors described above may allow stores to build biosensed profiles over time about individual customers and target advertisements or products directly to individuals.

The *limited controlability* of biosignals suggests that information that was once latently protected and hidden, such as identity, emotions, body temperature, activity level, or stress, are now open to being collected by a wide range of actors including mall management, store employees, and other customers. This information can be collected when one is simply in the presence of the biosensors, outside of the contexts of financial transactions or interpersonal interactions. Thus, it becomes easy to capture biosignals without customers' knowledge or consent, whether explicit or implicit. This means that customers may lose their ability to walk around a mall anonymously and they may lose plausible deniability about their emotional states.

The actors who receive the information for processing and dissemination are likely to expand. Because biosensed information is *precise yet ambiguous*, the data will likely be processed by and disseminated to third parties, such as data analytics companies or by the companies that create the technologies. These data are then interpreted and returned back to the end user. In many cases, the end users will not be the data subjects. For instance, a store may use a set of biosensors to collect data on its customers, and then use third-parties to analyze those data. This creates a new long-term and ongoing relationship between end users and the data recipients who process and disseminate the data. This stands in contrast to the traditional model where the company who sold a surveillance camera to a mall did not get full access to the recordings of every camera it sold. Furthermore, mall customers are unlikely to know of these data processing recipients or that their data is being sent to and analyzed by these recipients. The technical capabilities or new biosensors make it likely that this type of data processing, dissemination, and feedback can occur continuously rather than being tied to particular transactions, and can occur at an individual level rather than an aggregate one. Together, these shifts in the transmission principles suggest new types of privacy harms that may occur with the deployment of biosensing technologies in public spaces.

Biosensing: beyond information privacy

As highlighted above, in the examples of early photography and Google Glass, public fears about the ability to capture unsavory emotions or surreptitiously record individuals is particularly poignant in the case of biosensing technology. By using contextual integrity as a heuristic, the previous section further explored how remote biosensing in a public context may challenge and frustrate social and legal norms. Based upon the public context for biosensing, this chapter now considers how biosensing may require a new risk model for privacy extending beyond information privacy alone.

Drawing upon Tavani's (2008) categorization of four distinct types of privacy – (1) physical/accessibility, (2) decisional, (3) psychological/mental, and (4) informational – this section briefly examines the potential for biosensing in public to infringe on more than one meaning of privacy. Physical privacy, which focuses primarily on privacy as physical non-intrusion, may be considered the least relevant to public biosensing. Even in the cases where casual and unforced contact might be needed to mediate a sensor reading, there would be no mandatory or disruptive physical impact to the person.

Decisional privacy may be considered the freedom from interference affecting choices like education, health care, career, personal life, and beliefs (Floridi 2006). Biosensed information easily gives outsiders clues to individuals' reactions, attitudes towards an object or situation, or belies one's ambivalence about a situation. Gathering biosensed data creates an opportunity to interfere with or manipulate results. For instance, in Mexico, political campaigns have used biosignals like heart rate, facial coding, and neuro-feedback (collected from EEG data in the lab bacause, as of 2018, EEG data cannot be remotely or casually sensed) to assess voter attitudes and reactions to particular campaign material (Randall 2015). Collecting these data remotely in the wild to tailor advertising, including campaign material, is not far off. Smart-shelf technology has also been developed by companies like snack maker Mondelez International Inc. to collect real-time demographic and behavioral data about customers in order to present a customized ad experience that makes impulsive junk-food purchases even more tempting (Boulton 2013; Truth Labs 2014). These decisional interventions could be made to manipulate and interfere with countless areas of life, or be designed to prey upon our behavioral vulnerabilities like addictions.

The psychological or mental privacy to protect our intimate thoughts remains fairly straightforward for biosensed data. Scientists and evolutionary psychologists have studied why some cues are made public to assist us in public settings. For instance, tears are thought to be a subtle cue to instigate empathy from others within a few feet, but when biosensed data like emotions or internal physiological states are systematically recorded and analyzed, all signals become magnified beyond their original natural public scope. Further, inferences drawn from data individuals may not necessarily know or feel within themselves not only violates their inner most thoughts but, importantly, may circumvent our internal mental states altogether by presenting an algorithmic determination back to us, thus preventing our own emotional processes. The ways in which biosensing data in public could both intrude and interfere with our mental states and cause harm are numerous and present new challenges.

The informational meaning of privacy remains the most similar to other modern challenges toward limiting access and allowing control over one's personal information. Biosensing creates the potential for a greater amount of an entirely new data class of intimate information to be gleaned at a high rate from public encounters and kept for indiscriminate amounts of time. The ways in which information privacy may result in inappropriate flows of information and potential harms are numerous and vast.

Though Warren and Brandeis (1890) bring up emotions specifically in the call for a right to privacy, current legal and policy frameworks may not be able to fully cover privacy harms beyond information privacy. The ability to ascertain and make judgments (undoubtedly subjective and with bias) about internal emotional states using biosensing technology and using these inferences to interfere with decisional autonomy creates additional legal tensions relating to privacy.

Sensing and legal frameworks

The fuzziness around what type of data biosensing information may be categorized as emerging inferential power, along with the rapidly expanding potential for new remote technologies (that may be both inexpensive, discrete, and able to capture new unfamiliar data types) frustrate current legal frameworks. This chapter focuses on EU and US laws due to their respective prominence in global privacy law and contrasting approaches.

The EU General Data Protection Regulation (GDPR),[2] which superseded the Data Protection Directive[3] in May 2018, will likely prevent almost all public collection and processing of individuals' biosignals without explicit permission by the individual. "Personal data" is defined as "any information relating to an identified or identifiable natural person" where an identifiable person is one who can be directly or indirectly identified by referencing identifiers like name, ID number, location, or using one or more factors specific to that person – including physiological, genetic, mental, and social identity.[4] The regulation specifically calls out "biometric data" as a special category of personal data subject to heightened scrutiny.[5] Biometric data is defined by the regulation as "personal data resulting from the specific technical processing relating to physical, physiological, or behavioral characteristics of a natural person, which allow or confirm the unique identification of that natural person, such as facial images or dactyloscopic data."[6]

These definitions could be problematic because not all biosensing data may be identifiable or able to authenticate an individual but could still be revealing – a heightened heartrate signal can still be used to help make real-time assessments about a response or state even if it is not uniquely identifiable or combined with personal information. Though the ability to isolate biosignals from other indirect identifiers like location data is unlikely.

Further, it is possible for biosensing to frustrate EU restrictions on the collection of biosignals information by satisfying any number of exceptions listed under Article 9 of the GDPR. The law allows an exception if "processing relates to personal data which are manifestly made public by the data subject."[7] As discussed in this chapter, many of these signals are of limited control and outward facing, and thus may be considered "public" within the context of this law. Explicit consent from the data subject allows for the processing of personal data, and thus biosignals that are tied to identity. Given the familiar yet ambiguous and inferential potential of these signals, it is not unlikely that someone may consent to having their heart rate recorded without understanding how it may be used to make inferences about their health or emotional states. Consent also may be given without complete understanding of the ambiguous nature of these signals and the potential for improper categorization, like elevated heartrate inferring poor health state instead of genetic variability. Even with the ability to easily revoke consent, subjects may not fully understand the implications or inherent ambiguity of these signals in a way that motivates action to do so. The personal and "household use" of these data would also be permitted, which further underscores the importance of social norms in negotiating privacy in these unregulated social contexts.

Biosignals could also challenge broader clauses within the law, like processing that is "necessary for the purposes of preventative or occupational medicine," or for medical assessment. Many biosignals could lead to important biomedical inferences about individuals, and government, public health officials, or private healthcare providers may be able to use these justifications as means to wider collection and use. There may be additional exemptions such as those related to national security that open up other uses of biosignals. Even with the most comprehensive privacy laws to date, public biosignal collection may still pose privacy harms within the EU and spur further debates regarding the definition and classification of biosignals as a new class of sensitive data. The numerous ways in which GDPR may fully cover or leave opportunities for biosignal use is beyond the scope of this chapter, but will be important in future research as the policy and biosensing technologies mature.

Unlike the EU, the US has no single data protection law but has legislated on an *ad hoc* basis for particular sectors or circumstances. Privacy coverage under the Health Insurance Portability and Accountability Act (HIPAA)[8] would only be potentially extended to biosignals collected by organizations subject to the "privacy rule" – i.e., those defined as covered entities.[9] In situations where a patient presents their own collected biosignals to a healthcare professional, it is unclear at what point, if any, those data may be considered covered information. In general in the US, biosignals associated with wearable "wellness" devices are considered low risk and the Food and Drug Administration does not regulate the production and use of these devices. It could be possible that a law similar to HIPAA could regulate the exchange and storage of biosensing data when held by particular entities, but the coverage of potential data collectors and holders would be limited.

In the US, it could be possible for a privacy law like the Video Voyeurism Prevention Act of 2004[10] to prohibit the particular use of a type of technology. In response to the rise of cell phone cameras that were discrete enough to allow surreptitious image capture, notably around loose garments of unconsenting individuals, states and the federal government responded with video voyeurism laws to make this particular use of legal cameras illegal. A legal restriction on particular information gathering activities using otherwise legal instruments could theoretically be possible to prevent particular data collection from remote biosensing technologies.

Finally, within the US, the Federal Trade Commission (FTC) would likely be the strongest enforcer of privacy infringements through its mandate to prevent unfair and deceptive commercial practices (FTC 1980; FTC 1983). Cases brought by the FTC would be responsive to the undisclosed use, misuse, or insecure storage of biosensed data by commercial actors in the US. While cases may be reactive to harms caused to consumers, it may be possible for the FTC to issue guidance on responsible uses of biosensing technology and data similar to how the agency has done recently in regards to the internet of things (FTC 2015).

Conclusion: a consideration of remedies

In some ways, the ability to remotely collect emotional profiles or record others in public is similar to traditional photography and the tensions highlighted in the Google Glass debate. However, the unique nature of biosensed data as an emerging data class with particular attributes regarding meaning, integrity, and availability complicate existing social norms and privacy laws. As further demonstrated using contextual integrity, the ability to record remotely in public ushers in new tensions and highlights gaps in current legal structures that may not fully consider the ability to infringe upon decisional and emotional privacy in addition to expected information privacy concerns.

The possible remedies related to biosensing in public require additional research and debate. As discussed in this chapter, existing privacy laws and regulations in the EU and US contain gaps that could allow for harm by biosensing technologies and the collection, use, and sharing of associated data. However, broad restrictions on the use of personal information, like those in the EU, could help stem some harms linked to making inferences about an individual's mental or physical states, and additional regulation on the use of these technologies in non-identifying situations may be necessary. Better privacy protection could be afforded by restricting particular collection activities or biosensing technology uses, similar to the laws regulating video voyeurism with cameras. These laws could potentially act as a deterrent or create the opportunity for remedies of harms done. Additionally, protecting the storage and exchange of biosensing data in a HIPAA style law could prevent some harms by providing standards and protections for these data, but may be limited in the range of entities covered. Retroactive regulation for harms conducted by private industry or perhaps individuals will certainly be important as biosensing becomes more pervasive, but will necessarily be reactive instead of preventative.

Other potential remedies beyond law and policy may be important to adapting social norms. Personal modulation as a form of individual obfuscation may be possible and serve as a remedy against remote intrusion. Instrument jammers, data spoofers, or physical blocks may be worn to prevent some types of biosensing measurements. Masks may be used to prevent emotional analysis, and photography may be muddled by the use of reflective scarves that block images taken with flash (Access All Brands 2017). Though these are not geared specifically for biosensing, they represent opportunities individuals may be able to take in the absence of legal or social control. These remedies, and the roles they may have in developing social norms, should be explored in future work.

Community standards may be an option to complement social norms as a soft form of technology regulation. Social community standards may make particular actions using biosensing technology inappropriate and socially unacceptable due to peer pressure or public ridicule – or vice versa making evolving biosensing applications acceptable. Additionally, there may be an opportunity to develop technical standards into biosensing technology that allow a value to become embedded into the design – similar to efforts to make Do Not Track a standard for webpages online. This would likely be considered more of a technical standard setting effort as opposed to social community standard, but could offer a similar remedy.

Anticipating new privacy concerns in public spaces due to the rise in remote biosensing technologies, our use of the contextual integrity heuristic identifies several potential privacy concerns, and discussed potential remedies. This is just a first step in a broader conversation about privacy, ethics, and values – in the future, various ethical considerations may determine that other interests and values outweigh the privacy concerns of these technologies. Yet these conversations must occur actively (and proactively), among technologists, policymakers, and the general public so that technical, legal, and social decisions are made explicitly.

Acknowledgements

This work was funded and made possible by a generous grant from the Hewlett Foundation and Center for Long-Term Cybersecurity (CLTC) at the University of California, Berkeley. The authors would also like to thank the Berkeley BioSENSE Lab Group for their comments and feedback in the early stages of this manuscript proposal, the reviewers and book editors for their thoughtful and constructive feedback, and research assistant Matt Nagamine for his support and attention to detail.

This material is based upon work supported by the National Science Foundation Graduate Research Fellowship Program under Grant No. DGE1106400. Any opinions, findings, and conclusions or recommendations expressed in this material are those of the author(s) and do not necessarily reflect the views of the National Science Foundation.

Notes

1 See, e.g., *Pruneyard Shopping Center v. Robins*, 447 U.S. 74 (1980).
2 Regulation (EU) 2016/679, 2016 O.J. (L 119) 1.
3 Directive 95/46/EC, 1995 O.J. (L 281) 31 (EU).
4 Regulation (EU) 2016/679, art. 4, para. 1, 2016 O.J. (L 119) 1, 33.
5 Regulation (EU) 2016/679, art. 9, 2016 O.J. (L 119) 1, 38.
6 Regulation (EU) 2016/679, art. 4, para. 14, 2016 O.J. (L 119) 1, 33.
7 Regulation (EU) 2016/679, art. 9, para. 2(e), 2016 O.J. (L 119) 1, 38.
8 Health Insurance Portability and Accountability Act of 1996, Pub. L. No. 104–191, 1996 U.S.C.C.A.N. (110 Stat.) 1936 (codified at 42 U.S.C. §§ 1320d-1 to d-9 (2012)).
9 45 C.F.R. §§ 160, *et seq.*, 164.104 (2014).
10 18 U.S.C. § 1801 (2004).

References

Access All Brands. 2017. "ISHU: Anti-Flash Scarf." Accessed January 5, 2017. https://theishu.com/.

Adib, Fadel, Hongzi Mao, Zachary Kabelac, Dina Katabi, and Robert C. Miller. 2015. "Smart Homes that Monitor Breathing and Heart Rate." *Proceedings of the 33rd Annual ACM Conference on Human Factors in Computing Systems – CHI'* 15, 837–846.

Adler, Jerry. 2015. "Smile, Frown, Grimace and Grin – Your Facial Expression Is the Next Frontier in Big Data." *Smithsonian.com*, December. Accessed January 5, 2017. http://www.smithsonianmag.com/innovation/rana-el-kaliouby-ingenuity-awards-technology-180957204/.

Berisha, Visar, Shuai Wang, Amy LaCross, and Julie Liss. 2015. "Tracking Discourse Complexity Preceding Alzheimer's Disease Diagnosis: A Case Study Comparing the

Press Conferences of Presidents Ronald Reagan and George Herbert Walker Bush." *Journal of Alzheimer's Disease* 45(3): 959–963.

Bicchieri, Cristina. 2000. "Words and Deeds: A Focus Theory of Norms." In *Rationality, Rules, and Structure*, edited by Julian Nida-Rümelin and Wolfgang Spohn, 153–184. Dordrecht: Kluwer Academic Publishers.

Biotime Technology. 2014. "Finger on the Fly: Contactless 4 Fingerprint Capture in Less than 1 Second Now Available." *YouTube.com*, December 14. Accessed January 5, 2017. www.youtube.com/watch?v=lhAe36m4jQs.

Boulton, Clint. 2013. "Snackmaker Modernizes the Impulse Buy with Sensors, Analytics." *CIO Journal* (blog), *Wall Street Journal*, October 11. Accessed January 5, 2017. https://blogs.wsj.com/cio/2013/10/11/snackmaker-modernizes-the-impulse-buy-with-sensors-analytics/.

Bradfield, Ron, George Wright, George Burt, George Cairns, and Kees Van Der Heijden. 2005. "The Origins and Evolution of Scenario Techniques in Long Range Business Planning." *Futures* 37(8): 795–812.

Burns, Will. 2013. "Nine Business Applications for Google Glass." *Forbes*, February 27. Accessed January 5, 2017. www.forbes.com/sites/willburns/2013/02/27/nine-business-applications-for-google-glass-love-ideasicle/.

Cavoukian, Ann. 2011. "Privacy by Design: The 7 Foundational Principles." Accessed November 1, 2017. www.ipc.on.ca/wp-content/uploads/Resources/7foundationalprinciples.pdf.

Clifford, Stephanie, and Quentin Hardy. 2013. "Attention, Shoppers: Store Is Tracking Your Cell." *New York Times*, July 14. Accessed January 5, 2017. www.nytimes.com/2013/07/15/business/attention-shopper-stores-are-tracking-your-cell.html.

Clusmann, H., C. Schaller, and J. Schramm. 2001. "Fixed and Dilated Pupils after Trauma, Stroke, and Previous Intracranial Surgery: Management and Outcome." *Journal of Neurology, Neurosurgery & Psychiatry* 71(2): 175–181.

Computing Community Consortium. 2015. "Visioning Activity: Privacy by Design." Accessed January 6, 2017. http://cra.org/ccc/visioning/visioning-activities/2015-activities/privacy-by-design/.

Cooper, Charlie. 2014. "Facial Recognition Technology Used to Spot Genetic Disorders." *Independent*, June 23. Accessed January 5, 2017. http://www.independent.co.uk/news/science/facial-recognition-technology-used-to-spot-genetic-disorders-9558032.html.

Corkery, Michael. 2016. "Goodbye, Password. Banks Opt to Scan Fingers and Faces Instead." *New York Times*, June 21. Accessed January 5, 2017. www.nytimes.com/2016/06/22/business/dealbook/goodbye-password-banks-opt-to-scan-fingers-and-faces-instead.html.

Crawford, Kate, Jessa Lingel, and Tero Karppi. 2015. "Our Metrics, Ourselves: A Hundred Years of Self-Tracking from the Weight Scale to the Wrist Wearable Device." *European Journal of Cultural Studies* 18(4–5): 479–496.

Denning, Tamara, Zakariya Dehlawi, and Tadayoshi Kohno. 2014. "In Situ with Bystanders of Augmented Reality Glasses: Perspectives on Recording and Privacy-mediating Technologies." *Proceedings of the 32nd Annual ACM Conference on Human Factors in Computing Systems*, 2377–2386.

DiSalvo, David. 2013. "The Banning of Google Glass Begins (And They Aren't Even Available Yet)." *Forbes*, March 10. http://onforb.es/10rxJBl.

FTC (Federal Trade Commission). 1980. "FTC Policy Statement on Unfairness." Federal Trade Commission, December 17. Accessed January 5, 2017. www.ftc.gov/public-statements/1980/12/ftc-policy-statement-unfairness.

FTC (Federal Trade Commission). 1983. "FTC Policy Statement on Deception." Federal Trade Commission, October 14. Accessed January 5, 2017. www.ftc.gov/system/files/documents/public_statements/410531/831014deceptionstmt.pdf.

FTC (Federal Trade Commission). 2015. "Internet of things: Privacy & security in a connected world." Federal Trade Commission, January. Accessed January 5, 2017. www.ftc.gov/system/files/documents/reports/federal-trade-commission-staff-report-november-2013-workshop-entitled-internet-things-privacy/150127iotrpt.pdf.

Floridi, Luciano. 2006. "Four Challenges for a Theory of Informational Privacy." *Ethics and Information Technology* 8(3): 109–119.

Food and Drug Administration. 2016. "General Wellness: Policy for Low Risk Devices – Guidance for Industry and Food and Drug Administration Staff." July 29. Accessed January 5, 2017. https://www.fda.gov/downloads/MedicalDevices/DeviceRegulationandGuidance/GuidanceDocuments/UCM429674.pdf.

Friedman, Batya, Peter H. Kahn, Jr., Jennifer Hagman, Rachel L. Severson, and Brian Gill. 2006. "The Watcher and the Watched: Social Judgments about Privacy in a Public Place." *Human-Computer Interaction* 21(2): 235–272.

Guston, David H. 2014. "Understanding 'Anticipatory Governance.'" *Social Studies of Science* 44(2): 218–242.

Hewitt, John. 2013. "MIT Researchers Measure Your Pulse, Detect Heart Abnormalities with Smartphone Camera." *ExtremeTech*, June 21. Accessed January 5, 2017. www.extremetech.com/computing/159309-mit-researchers-measure-your-pulse-detect-heart-abnormalities-with-smartphone-camera.

Hill, Kashmir. 2012. "Sergey Brin's Favorite Google Glass Feature." *Forbes*, September 11. Accessed January 5, 2017. http://onforb.es/RIXia6.

Humphreys, Lee. 2005. "Cellphones in Public: Social Interactions in a Wireless Era." *New Media & Society* 7(6): 810–833.

Humphreys, Lee. 2010. "Mobile Social Networks and Urban Public Space." *New Media & Society* 12(5): 763–778.

ISO (International Organization for Standardization). 2017. "ISO/IEC 27001 Family – Information Security Management Systems." Accessed January 6, 2017. http://www.iso.org/iso/iso27001.

Introna, Lucas, and Helen Nissenbaum. 2009. "Facial Recognition Technology: A Survey of Policy and Implementation Issues." The Center for Catastrophe Preparedness and Response, New York University, July 22. Accessed January 5, 2017. https://ssrn.com/abstract=1437730.

Klosowski, Thorin. 2013. "How Stores Manipulate Your Senses So You Spend More Money." *Lifehacker*, April 18. Accessed January 6, 2017. http://lifehacker.com/how-stores-manipulate-your-senses-so-you-spend-more-mon-475987594.

Latvala, Antti, Ralf Kuja-Halkola, Catarina Almqvist, Henrik Larsson, and Paul Lichtenstein. 2015. "A Longitudinal Study of Resting Heart Rate and Violent Criminality in more than 700000 Men." *JAMA Psychiatry* 72(10): 971–978.

Li, Xiaobai, Jie Chen, Guoying Zhao, and Matti Pietikäinen. 2014. "Remote Heart Rate Measurement from Face Videos under Realistic Situations." *Proceedings of the IEEE Computer Society Conference on Computer Vision and Pattern Recognition*, 4264–4271.

Li, Xiaobai, Xiaopeng Hong, Antti Moilanen, Xiaohua Huang, Tomas Pfister, Guoying Zhao, and Matti Pietikäinen. 2015. "Towards Reading Hidden Emotions: A Comparative Study of Spontaneous Micro-Expression Spotting and Recognition Methods.", *IEEE Transactions on Affective Computing PP* (99): 1–13.

Licht, Amir N. 2008. "Social Norms and the Law: Why Peoples Obey the Law." *Review of Law and Economics* 4(3): 715–750.

Lindsay, David. 2000. "The Kodak Camera Starts a Craze." PBS. Accessed January 5, 2017. http://www.pbs.org/wgbh//amex/eastman/peopleevents/pande13.html.

MacGillivray, Tom, Manuel Trucco, James Cameron, Baljean Dhillon, Graeme Houston, and Edwin van Beek. 2014. "Retinal Imaging as a Source of Biomarkers for Diagnosis, Characterisation and Prognosis of Chronic Illness or Long-Term Conditions." *The British Journal of Radiology* 87: 1–16.

Madden, Mary, and Lee Rainie. 2015. "Americans' Attitudes about Privacy, Security and Surveillance." *Pew Research Center*, May 20. Accessed January 5, 2017. www.pewinternet.org/2015/05/20/americans-attitudes-about-privacy-security-and-surveillance/.

Massimi, Michael, Khai N. Truong, David Dearman, and Gillian R. Hayes. 2010. "Understanding Recording Technologies in Everyday Life." *IEEE Pervasive Computing* 9(3): 64–71.

Mullan, Patrick, Christoph M. Kanzler, Benedikt Lorch, Lea Schroeder, Ludwig Winkler, Larissa Laich, Frederik Riedel, et al. 2015. "Unobtrusive Heart Rate Estimation during Physical Exercise Using Photoplethysmographic and Acceleration Data." *Proceedings of the Annual International Conference of the IEEE Engineering in Medicine and Biology Society, EMBS 2015-November*, 6114–6117.

Mulligan, Deirdre K., and Jennifer King. 2011. "Bridging the Gap between Privacy and Design." *University of Pennsylvania Journal of Constitutional Law*, 989–1034.

NIST (National Institute of Standards and Technology). 2014. *Framework for Improving Critical Infrastructure Cybersecurity*. NIST. Accessed January 5, 2017. www.nist.gov/sites/default/files/documents/cyberframework/cybersecurity-framework-021214.pdf.

Nguyen, David H., Alfred Kobsa, and Gillian R. Hayes. 2008. "An Empirical Investigation of Concerns of Everyday Tracking and Recording Technologies." *Proceedings of the 10th international conference on Ubiquitous computing (UbiComp '08)*, 182–191.

Nissenbaum, Helen. 2010. *Privacy in Context*. Stanford: Stanford University Press.

Palca, Joe. 2014. "Chemist Turns Software Developer after Son's Cancer Diagnosis." *NPR*, May 6. Accessed January 5, 2017. http://www.npr.org/sections/health-shots/2014/05/06/309003098/chemist-turns-software-developer-after-sons-cancer-diagnosis.

Randall, Kevin. 2015. "Neuropolitics, Where Campaigns Try to Read Your Mind." *New York Times*, November 3. Accessed January 5, 2017. www.nytimes.com/2015/11/04/world/americas/neuropolitics-where-campaigns-try-to-read-your-mind.html

Räty, Tomi D. 2010. "Survey on Contemporary Remote Surveillance Systems for Public Safety." *IEEE Transactions on Systems, Man, and Cybernetics, Part C (Applications and Reviews)* 40(5): 493–515.

Rubinstein, Ira S. 2011. "Regulating Privacy by Design." *Berkeley Technology Law Journal* 26: 1409–1456.

Sarvas, Risto, and David M. Frohlich, eds. 2011. "The Kodak Path (ca. 1888–1990s)." In *From Snapshots to Social Media – The Changing Picture of Domestic Photography*, 47–82. London: Springer London.

Scigiene Corporation. n.d. "Infrared Thermometers." Accessed January 5, 2017. www.scigiene.com/pdfs/Infrared%20Thermometers.pdf.

Solove, Daniel. 2006. "A Taxonomy of Privacy." *University of Pennsylvania Law Review* 154(3): 447–560.

Solove, Daniel, and Paul M. Schwartz. 2015. *Information Privacy Law*. New York: Wolters Kluwer.

Surden, Harry. 2007. "Structural Rights in Privacy." *Southern Methodist University Law Review* 60: 1605–1629.

Tavani, Herman T. 2008. "Informational Privacy: Concepts, Theories, and Controversies." In *The Handbook of Information and Computer Ethics*, edited by Kenneth E. Himma and Herman T. Tavani, 131–164. Hoboken, NJ: John Wiley & Sons, Inc.

Tene, Omer, and Jules Polonetsky. 2013. "A Theory of Creepy: Technology, Privacy and Shifting Social Norms." *Yale Journal of Law and Technology* 16(1): 59–102.

Thier, Dave. 2013. "The Potentially World-Changing New Feature in Google's Project Glass." *Forbes*, February 20. Accessed January 5, 2017. http://onforb.es/11VjbyK.

Trottier, Daniel. 2014. "Crowdsourcing CCTV Surveillance on the Internet." *Information, Communication & Society* 17(5): 609–626.

Truth Labs. 2014. "Mondelez Smart Shelf." *Vimeo.com*. Accessed January 5, 2017. https://vimeo.com/96148396.

Wang, Siying, Antje Pohl, Timo Jaeschke, Michael Czaplik, Marcus Kony, Steffen Leonhardt, and Nils Pohl. 2015. "A Novel Ultra-Wideband 80 GHz FMCW Radar System for Contactless Monitoring of Vital Signs." *Proceedings of the Annual International Conference of the IEEE Engineering in Medicine and Biology Society*, 4978–4981.

Warren, Samuel D., and Louis D. Brandeis. 1890. "The Right to Privacy." *Harvard Law Review* 4(5): 193–220.

West, Nancy Martha. 2000. *Kodak and the Lens of Nostalgia*. Charlottesville: University of Virginia Press.

Wong, Richmond Y., and Deirdre K. Mulligan. 2016. "When a Product Is Still Fictional: Anticipating and Speculating Futures through Concept Videos." *Proceedings of the 2016 ACM Conference on Designing Interactive Systems*, 121–133.

Chapter 6

Adverse detection

The promise and peril of body-worn cameras

Michael A. Katell

Introduction

Raina McDonald was in the front yard of her home when two police officers emerged from a squad car on their way to check out a party going on at her next-door neighbor's house. The officers were wearing body-worn video cameras. After officers confronted one of the party guests, a scuffle broke out and the party guest was subdued and arrested by the officers. Raina later heard that some video of the incident was posted on the police department's YouTube channel and that Raina could be clearly seen in the background. She didn't give it much more thought.

Raina is African American and her neighborhood is predominantly African American as well. Police activity is fairly common there, which is one reason Raina decided, a few months later, to submit a rental application for a home in a different part of town. When her application was turned down the property manager told her that, despite her good credit and positive references, the company's background check provider reported that Raina had a low ClearLiving^{TM} score. He couldn't tell Raina anything about the score or why hers was low because he had no idea. All he knew was that it was the management service's policy to reject rental applicants with a score in her range. Raina contacted the company who provided the background report and asked for more information about her score. In response to her inquiry, she received a polite reply explaining that they couldn't tell Raina how their scores are derived and what goes into them because the technology and data sources are proprietary and confidential.

The body-worn camera video from the incident next to her home was partly to blame for Raina's disappointing ClearLiving^{TM} score. Data from police patrol recorders, and other records of police activity, are considered public records where Raina lives, generally available to anyone who requests them. The local police department also has a policy of uploading video footage to their YouTube channel after violent encounters between police and citizens. Data mining companies use automated crawlers to routinely access the government databases where police records are stored. The data miners also scrape content from video sharing sites, including police department YouTube channels. Using reidentification techniques – including facial recognition and data triangulation – the data miners connect video images with the identities of people in the videos. By combining reidentified police data with the thousands of other data points about Raina that can be gleaned from CCTV footage, government databases, credit card processors, social media platforms, retail websites, mobile apps, insurance companies, and other data sources, a wealth of documentation about Raina's life is available to generate a digital profile of her that can be converted into quantified reputational assessments like the score used by the property management company in this account.

Raina's image, captured in a public space and then incorporated into her digital profile, resulted in her being categorized as someone observed at or near a police incident. The scoring algorithm that produced her ClearLiving™ score has been trained to identify patterns and find correlations between facts like physical proximity to crime and the potential risk of criminality. Because the data is gathered and processed by intelligent software with little human intervention, the company who markets the ClearLiving™ score probably couldn't say exactly how Raina's score was derived even if they wanted to. The process is opaque to the data processor, the data consumer, and the data subject.

Raina McDonald is not a real person. The names of the data products and systems described above are also fictitious. However, while speculative, the story about Raina describes the convergence of three information trends: the increasingly common practice of recording and documenting people's public and private lives, the increasing quantities of government data about people being created and shared with multiple data consumers, and the business activity of acquiring and analyzing data to make people completely knowable and predictable to decision makers. All of the technical artifacts, police practices, policy decisions, business operations, and market forces described in Raina's story already exist. While a certain amount of speculation is required in order to interrogate the specific, proprietary practices of the data industry, the availability of data, the state of data analytics research and innovation, and increasing market demand for predictive insights, leave us with reason to believe that scenarios like the one describing Raina McDonald's rental application can and will occur.

Aims

The goal of this chapter is to sketch a causal chain that links the expansion of surveillance practices in public spaces with downstream effects on human flourishing. In particular, this chapter highlights the tension between the high hopes for a particular technical artifact – the police body-worn video camera – and its relationship to the world of big data, with a particular focus on the so-called reputation economy. I also surface a thread of bias and discrimination against marginalized segments of society due to the nature of state surveillance that begins at the point of data collection about people and ends in the potentially life-altering decisions of social and economic gatekeepers.

I begin this chapter with a look at the police body-worn camera as a relatively novel component of state surveillance. I then go on to suggest that the data produced by the police, including body-worn camera footage, is inherently biased against some members of society by the nature of police work and the social construction of suspicion. This leads to a discussion of a concept I call "adverse detection," which ties together the loss of privacy in public through increasing surveillance, open government records, reidentification of data subjects, and the increasingly quantified nature of reputation. I argue that adverse detection is a potentially marginalizing byproduct of being observed and recorded in public. I also postulate how police video could make its way from government databases into the data warehouses of commercial data brokers, leading to a discussion of big data analytics and its expression in the burgeoning reputation industry. I bring together the questions I previously raised about the promise of police body-worn cameras, the undervaluing of their potential harm to the experience of public space, and the potentially negative influence on the life opportunities of those who are judged by their digital reputations. I close with a series of recommendations including calling for a closer examination of public disclosure laws and addressing the black-box secrecy of reputational decision-making.

Surveillance

Surveillance, as a systematic and routine practice by government officials, is a longstanding phenomenon whose aims include influencing, managing, protecting and disciplining a given population (Lyon 2007; Foucault 1977; Goos et al. 2015). Scholars have been considering the transformative nature of digitally mediated, pervasive surveillance for decades as long-held perceptions of being anonymous in public have fallen away due to the relentless logging and analysis of our movements, communications, interactions, and transactions. A very brief overview of the surveillance literature includes Haggerty and Ericson (2000), who describe the dehumanizing web of logging and classifying technologies that permeates modern life as a single, dominating phenomenon they label the "surveillant assemblage" (614). Lyon (2003) frames the societal effects of pervasive surveillance as "social sorting," the creation and enforcement of social and economic hierarchies whose effects and raise important questions about social justice and the distribution of power. Nissenbaum (1998) explores the under-valued and frequently dismissed concept of a right to privacy in public. Nissenbaum rejects the prevailing assumption that every appearance and action outside of one's domicile should be considered unassailably public and therefore open to unlimited sensing, recording, storing, and analysis. The argument that people should have some privacy expectations in public spaces sharply contrasts with policymaking and court decisions in the United States which rarely assert such rights.

Electronic surveillance of public space has become increasingly pervasive in much of the world due to advances in the development of light-weight, relatively inexpensive audio and video capture technologies whose affordances enable their data to flow smoothly into existing networks without the intervention of experts or much expense (Goos et al. 2015). Government and private institutions actively participate in public surveillance in order to pursue public safety, social control, or mercantile goals. But the ease of capturing, publishing and sharing amateur videos using consumer technology and popular internet sites has also enabled ordinary people to participate in public surveillance and to engage in new forms of political activism. In particular, the relative ease of capturing and sharing images of law enforcement activity has provided a new visibility (literally) to decades-old controversies about police accountability and violence. Video footage has been central to controversies surrounding deadly encounters involving African Americans and police officers in the United States (NY Daily News 2016; BBC News 2015). The potential for video evidence to shed light on contested narratives about police violence has fuelled the public's enthusiasm to equip police officers with body-worn cameras.

The promise of police body-worn cameras

As a recent entrant to the apparatus of public surveillance, the police body-worn camera is being rapidly adopted around the world. Existing or planned implementations of the technology in the United States are so pervasive as to be nearly total (Joh 2016). There is a perception that increasing the number of police–public encounters that are captured on video will result in revealing the details of police conduct and hold officers accountable when they choose to use force. This has resulted in body-worn camera use by the police being championed by a wide range of government officials and civil rights advocates, and even some (but not all) police officers (Jennings, Fridell, and Lynch 2014; Stanley 2013). The enthusiasm of civil rights advocates, who have been skeptical of state surveillance practices generally, is unique to this particular artifact and stems from hopes that body-worn camera use by the police will dramatically improve the acuity of community oversight over law enforcement practices. The stated hopes for body-worn cameras include a "civilizing effect" on both police officers and citizens that may follow from the panoptic effects of inserting a camera into their interactions (White 2014, 35).

Initial research seems to support these claimed benefits. In a study frequently cited by scholars investigating the social and policy issues surrounding body-worn cameras (e.g. White 2014; Newell 2017; Miller, Toliver, and Police Executive Research Forum 2014; Jennings, Fridell, and Lynch 2015), a randomized control trial of body-worn camera use by police in Rialto, California found a correlation between the use of the cameras and a reduction in police use-of-force incidents (Ariel, Farrar, and Sutherland 2014). However, despite widespread celebration of these findings by policymakers, industry leaders, and the press (Weise 2016), there is contrasting data, including the results of a subsequent study conducted by some of the same authors involved in the Rialto research. This later study contradicts some of the findings in the 2014 article by Ariel et al., indicating that there may actually be little or no change in police conduct where body-worn cameras are used, and that violence may actually increase against officers equipped with cameras (Ariel et al. 2016).

Additional doubts about the value of body-worn cameras were introduced by a video presentation published in the *New York Times* that demonstrates, using actors and multiple camera angles, the difficulties of accurately interpreting body-worn camera video as reliable evidence due to the close and narrow perspective of the cameras. Video producers demonstrated how a subject's non-aggressive but active behavior (such as dancing) can be mistaken as threatening behavior when viewed as body-worn camera video, casting doubt on the reliability of such footage for accurately reconstructing the details of an encounter (Williams et al. 2016). This demonstration adds additional ambiguity to the empirical findings mentioned above. While body-worn cameras may increase the visibility of police work, the question is still open as to whether or not such visibility will lead to the changes desired by advocates for greater police accountability. Indeed, commentators have pointed out that, despite the existence of video evidence that would seem to implicate police officers in several high-profile deaths, there have been few successful prosecutions of the officers involved ("Considering Police Body Cameras" 2015). As these doubts permeate the debate about body-worn camera use, a largely overlooked topic is the contribution to expanded state surveillance of public spaces enabled by these devices.

Police officers go everywhere in our towns and cities: the streets, private homes and businesses, hospitals, parks, schools, churches, and so on. The expansive reach of the police – when joined with recording devices worn on their person – produces a particularly rich source of data that is valuable to both law enforcement and anyone else who can gain access to it. Body-worn cameras can gather observations in obscure locations that stationary devices are not currently targeting; including locations that, while presumed to be public, may be viewed as *personal* if not private spaces, like the front steps of Raina McDonald's home on her residential street. Unlike surveillance conducted through unaided human observation, the type of surveillance enabled by body-worn cameras takes lasting notice of every single person in the shot. Newell (2017) describes the condition of involuntarily entering the public record as an identifiable video subject, along with the associated damage to one's privacy interests, as "collateral visibility" (10).

Legal doubts about the legitimacy of asserting a right to privacy in public coupled with government records policies that favor open access, join together to create an environment in which body-worn cameras raise the specter of collateral visibility for anyone appearing in a public place.

The inherent bias of police-generated surveillance data

Data collected in the normal course of police work is not neutral data. It is essentially selective because law enforcement is essentially selective. Police forces have neither the resources nor the will to detect and enforce every violation of the law that occurs in their jurisdictions (Cahill 2007, 820). As a result, one of the primary features of law enforcement includes the construction of suspicion by police officers based on their observations of the public and their discretion in choosing when to act. Additionally, a proportion of police work is driven by complaints that are reported by others to emergency personnel. Here again, not every violation of the law will come to notice because there is not always a witness to every incident or someone willing to come forward to report it, which also contributes to the selective nature of enforcement.

This general selectivity in police data collection is tainted by evidence suggesting that police activity is aimed more acutely at members of specific racial and ethnic groups. In a particularly stark example, a large scale study of traffic stops conducted in the United States showed that African American and Latino drivers were stopped at disproportionately higher rates than White drivers (Pierson et al. 2017). They were also much more likely to be searched during those stops than White motorists, yet searches of cars driven by White motorists were more likely to yield contraband than searches of non-White drivers. Racial profiling appears to have been a factor in these traffic stops, and yet that profiling did not accurately assess potential wrongdoing (Pierson et al. 2017). Traffic stops are not the only likely locus of biased policing. Survey findings show that illicit drug use and weapons violations are fairly uniform across racial divides. Yet statistically, African American men in the United States are much more likely to be arrested and charged for drug and weapons crimes than White men (Snyder 2012; Fellner 2009; Council on Crime and Justice 2007). Wherever disparate treatment by the police of one population over another is in effect, any data collected or produced will reflect the same (or similar) disproportionate distributions across those populations.

There are wide-ranging theories about what causes the over-policing of some populations, including theories about the social construction of suspicion. The construction of suspicion is particularly salient in the observation of unknown people in public spaces. As Norris and Armstrong (1999) discovered in their research on British security personnel engaged in CCTV surveillance practices, while there may be obvious grounds to suspect particular individuals (e.g., those seen visibly engaging in apparently criminal activity), the construction of suspicion often occurs for "no obvious reason." Suspicion frequently emerges as a hunch rather than due to an observation of explicit wrongdoing.

Many such hunches seemed to fall disproportionately on young, non-White men, suggesting deep-seated inclinations, or "value prejudices," on the part of the observers (Norris and Armstrong 1999, 163–164). In an attempt to understand the decision-making processes of surveillance operators, Smith (2012) describes one effect of routinely witnessing distressing events as "damaged subjectivity," which is the formation of a pessimistic view of human nature stemming from one's negative experiences (113). Police officers, who routinely experience the most chaotic and traumatic events in our communities, may form suspicions from the combined effects of value prejudice and damaged subjectivity resulting in a generalized suspicion targeted to entire segments of society. Suspicions built on bias will tend to create data whose contours follow the shape of those suspicions. The potential result is the overproduction of police data about members of marginalized populations.

In a case like that of Raina McDonald and her neighbors, we might be justified in wondering if racial bias played a role in drawing the attention of the police, and whether a similar situation in which the subjects were not African American would have gone differently. Such questions go to the heart of questioning the basic justice of public surveillance by the police and the presumed neutrality of the data produced by their encounters.

The dimension of (some) public spaces

The potential overproduction of police data about marginalized populations is not limited to the primary targets of police activity (those suspected of crimes); it is also indicated by the inclusion of *bystanders* to police activity, whose bodily images are frequently collected as a byproduct of video recording by the police. This implicates public spaces as a venue for biased surveillance practices. However, not all public spaces are identical. Due to the legacy of discriminatory housing policies in the United States (i.e. "redlining"), the effects of which remain evident into the 21st century, members of non-White communities frequently live in close proximity to each other and in specific geographic areas (powell 2008). The emphasis on enforcement actions on specific demographic groups is frequently tied to a focus on the geographies in which those groups are concentrated. The demography of a particular public is thereby linked to the geography of public space. As a result, the most likely bystanders to police contact with marginalized people are those who share demographic affinities – typically other members of the same marginalized group. Any seemingly random or coincidental over-inclusion of African Americans and other non-White bystanders in body-worn camera video is not random at all. It emerges from ongoing police practices that bring law enforcement to bear disproportionately on particular groups based on the demographics of their neighborhoods, the origins of which may trace their roots to official policies of discrimination.

Adverse detection

The assumption of being anonymous while going about one's business in public space can no longer be taken for granted. CCTV cameras, cell phone location trackers, automated license plate readers, automobile toll transponders, and numerous other vectors of observation and tracking are transforming public spaces into venues of constant surveillance (Blanchette and Johnson 2002; Nissenbaum 1998; Doyle and Veranas 2014). Police body-worn cameras in particular make an important contribution to this assemblage because of the authority granted to the police to go almost anywhere and the potentially biased choices they make in doing so.

In the past, police officers without cameras and other sensors might have made a brief mental note of people proximate to but not directly connected to an incident, particularly if it occurred in a public place. They were unlikely to create lasting records about bystanders. With the increasing frequency of video recording by police of their scenes of action, bystanders, like Raina McDonald, join other innocents in becoming collaterally visible without consent. Merely collecting data, however, is not the only problem. The information industry actively seeks out new collections of data about people from any source as it becomes available. And one of the ways they hope to extract value from such data is in marketing of reputational ratings and scores. Wherever data sharing is not restricted by law, data collection by state actors, such as police officers, can become part of this information economy.

I refer to the risk of having one's life documented *and analyzed* through data practices that can lead to negative outcomes for the data subject as "adverse detection." Adverse detection extends Newell's (2017) concept of "collateral visibility" and joins it to aspects of Allen's (2012) "information injustice," which describes the harm of being entered into a data set as a consequence of an unjust act, such as a racially biased arrest. Adverse detection goes beyond the effects of merely being recorded, identified, and publicized after being captured by surveillance to include the *processing* of the information into data points for quantified judgements. Adverse detection concerns not only the unjust collection of data about people and their identification without their consent, but includes deleterious judgements – like a rejected rental application – occurring on the basis of that data.

Public space and open government records policy

There is an inclination in the United States to assert the people's sovereignty over national and local governments by requiring public access to government records. Most U.S. states have expansive public disclosure laws which mandate open access to government records. Body-worn camera video, as a government record, may be subject to these laws. The notion of "the public" has expanded since many public disclosure laws were drafted and now includes commercial data brokers who have been mining government records since the 1970s for targeting direct marketing and advertising campaigns (Solove 2004). Meanwhile, due to the overall expansion of government and the relative ease of creating and storing digital records, the quantity of records has grown dramatically. Among the additions to these collections are the digital detritus of routine surveillance of public space from a multitude of devices including traffic cameras on city streets, toll transponder sensors on bridges and highways, and of course, patrol audio and video data collected by police officers.

I predict that body-worn camera video is poised to join other records of public surveillance as an input into the rich tapestry of personal data accumulated and analyzed by the data industry. Admittedly, a clear link between police video data and its use in commercial practices is difficult to positively assert at this time. Research on this topic is challenging because of the proprietary nature of the data industry's methods and the legal protections afforded by trade secrets law in the United States. We may be able to detect, in some cases at least, who has accessed or requested particular information artifacts from government sources, but established business norms and resistance to official inquiries prevent us from gaining deeper insights into exactly how the data is being used in commercial applications (Rockefeller 2013).

As for the legal processes involved, in the United States there is a mix of hurdles and pathways to linking body-worn camera video to commercial data mining. Almost every state applies some restriction on the public disclosure of body-worn camera video, but in many cases the restrictions are quite narrow opening the door to at least some data mining activities. For example, the U.S. states of Connecticut, Florida, Louisiana, Missouri, Wisconsin, and Utah are among those that restrict public access to body-worn camera video only under certain conditions, such as if the data pertains to an ongoing investigation or its disclosure could result in the identification of a confidential informant (see Newell 2017 for a discussion of some of these laws). In contrast, the states of Alaska, California, Colorado, North Carolina, and Washington are among those whose access restrictions are more comprehensive, providing fewer avenues of disclosure (Urban Institute 2016; see also Newell 2017). But the devil is in the details of all of these laws, which bump up against long-held U.S. traditions of free speech, unhindered journalism, and government transparency. For example, the State of Washington has a strongly worded Public Records Act that only allows exemptions for information that meets the standard of being either "highly offensive to a reasonable person" or "is not of legitimate concern to the public."[1] In the past, when challenged in court, the latter "legitimate concern" clause of the Act has proven difficult to overcome (see e.g., Hundtofte v. Encarnacion 2014). Washington passed a law in 2016 broadly exempting body-worn camera recordings from public disclosure and included language in the law specifically stating that the recordings presumptively meet the "highly offensive" standard.[2] However, the legislation is notably silent about whether body-worn camera video is "not of legitimate concern to the public." Consequently, Washington's body-camera video exemption may be vulnerable to a legal challenge.

The patchwork nature of laws concerning the disclosure of body-worn camera video points towards at least some of this data being available to data brokers. Even existing prohibitions may not withstand the legal challenges that are routinely issued by news organizations and the data industry when policy makers seek to restrict access to some government records (e.g., Florida Star v. B. J. F. 1989; Sorrell v. IMS Health Inc. 2011). This leaves body-worn camera data potentially vulnerable to disclosure in the future even where it is currently restricted by law.

Identification

The technical feasibility of identifying the people contained in government records is an increasingly trivial matter even when government bureaus endeavor to strip out or redact personally identifiable information. Researchers have demonstrated the vulnerability of seemingly "anonymized" textual data to techniques of reidentification (Sweeney 2015; Barbaro and Zeller 2006). Meanwhile, advances in biometric identification technologies are taking place in both governmental and commercial contexts (Garvie, Bedoya, and Frankle 2016; Schmidt 2011). In particular, the use of facial recognition by law enforcement is expected to increase (Williams 2015). Hoping to profit from these developments, the primary manufacturer of body-worn cameras used by police departments in the United States is preparing a slate of analytics products that will use the cameras to identify people in public space in real time, picking out possible suspects and reporting their histories to police officers (Weise 2016). Wherever police video recordings become publically available, the increasing ease of identifying the subjects of those videos – suspect and bystander alike – makes such records very attractive to a variety of interests whose goals go well beyond the purposes for which the records were originally created.

In an effort to protect video subjects before releasing video to the public, many police forces are relying on video editing techniques to redact the faces of some subjects, but the work is time-consuming and cost-prohibitive (Hinds 2013; Sullivan and Miletich 2014). Even if meaningful, affordable redaction could be feasibly accomplished to protect their identities, the efficacy of current approaches to prevent (re)identification of video subjects could be fleeting at best. Federal law enforcement agencies are actively encouraging research on other forms of identification enhancement including gait analysis (identifying people based on their distinctive movements) and tattoo recognition, among others (Chen 2014; Maass and Mackey 2016). It is probably only a matter of time before these, or other biometric identification techniques are widely available.

The technical feasibility of identifying people portrayed in police video adds another dimension to the impact on public space wrought by police body-worn cameras. It is not only law enforcement agencies that are motivated to identify video subjects. The data industry, in its quest for ever more data about people to analyze and monetize, has much to gain from identifying the subjects of police video. A primary reason to be concerned about this is that police videos typically portray negative events including people having "the worst day of their lives" (Hinds 2013). In keeping with the widespread belief that events that take place out of doors cannot be presumed to be private, policy makers have defended police patrol recordings as presumably public as well (see e.g., Ferguson 2014). Combined with concerns about the inherent bias of police generated data, identification of body-worn video subjects through data mining activities could have negative consequences for anyone who spends any part of their life in public space.

The reputation economy

An emerging aspect of the information industry is focused on harnessing the power of automated algorithmic analysis (aka "machine learning") and the wealth of detailed data about virtually everyone[3] to create data products to predict human behavior. This emerging industry, which could be characterized as the marriage of direct marketing, consumer analysis, and financial credit reporting, has been labeled "the reputation economy" by Fertik and Thompson in their book of the same name (Fertik and Thompson 2015). In the reputation economy consumer scores and quantified categories are marketed to decision makers "to make major decisions about your life, such as whether to hire, insure, or even date you" (5). The reputation economy is built on the work of the thousands of modern data brokers who have been developing profiling methodologies for years to target consumers, first for internet advertising and direct marketing, and now to construct quantified reputational analyses about specific individuals. There is a lot of data out there to work with. One of the largest data brokers, Acxiom, claims to have an average of 1,500 pieces of information on more than 200 million Americans (Boutin 2016).

In the reputation economy data brokers distill rich data about human behavior into simple identifiers and benchmarks claiming to predict virtually anything, from the likelihood of a consumer switching to a new cell phone provider to the risk of her impending death (Singer 2014). There are serious concerns that the processes that create consumer scores and categories can surface sensitive information, sharply limit data subjects' opportunities, or put them at risk of exploitation. In the U.S. Congress, a committee studying consumer scoring harshly criticized what they considered stigmatizing reputation classifications such as "rural and barely making it" and "tough start: young single parents" (Rockefeller 2013). The risk of consumer scoring is highlighted by a number of regulatory actions including one in which a data broker faced sanctions for selling a contact list of thousands of people whose reputational assessments showed them to be both elderly and enthusiastic sweepstakes players. The list buyers were scam artists who stole over $100 million from people on the list (Boutin 2016).

Government records are crucial ingredients in the emerging reputation economy. When police surveillance data is part of such records, the bias in the data potentially echoes in the analyses of the reputation industry. For example, the context in which a record is created is as much a part of the character of the record as its content. Just as a credit card company may interpret paying for marriage counseling as a signal of financial instability and raise a customer's interest rate as a result (Pasquale 2015, 14), being identified in police surveillance data sends a strong signal about the data subject that could result in negative decisions being made about her. The reputation economy intersects with the increasing visibility of public spaces through surveillance artifacts such as police body-worn cameras. Living in a geographic location where police activity is a routine occurrence can lead to the creation of disadvantageous reputational records, which makes appearing in public potentially hazardous to a data subject's life chances whether or not she has anything to do with the interests of law enforcement. In the story of Raina McDonald we see an example of someone who should fall under the broad category of having "nothing to hide", and yet the choices of her neighbors and the practices of her local police joined to the exigencies of the reputation economy can make her appear undesirable as a prospective renter just the same.

Being unfairly judged based on data is a potential problem for everyone, but for some it is even more important. Reputation acts as a gatekeeping device, standing between those seeking to participate in economic and civic life, and the set of opportunities that enable that participation. All else being equal, if one job applicant has a higher reputational assessment from a trusted source or an influencer than a second applicant, the first applicant is likely to prevail. However, as argued by Julia Fleming (2004), all else is *not* equal. For members of unprivileged classes and groups, who do not enjoy access to as many informal social and economic networks as members of dominant groups, reputation is crucially important. A member of a non-dominant group, lacking the advantages of informal networks of privilege, may have *only* her reputation (76). Reputation is arguably more important to those without social power, but it also more vulnerable due to institutional biases, as well as how, where, and by whom the information is gathered. As I argue in this chapter, body-worn camera video, like most other data from police work, is likely to reflect institutional and societal biases. The potential for body-worn cameras to participate in the reputation economy, and the capability of its products to provide potentially discriminatory signals to the highest bidder, raises the specter of digital disenfranchisement for those who are most vulnerable to reputational judgement.

Reputation and social sorting

While it might be argued that the urge to categorize people and assess risk reflects natural human tendencies, the fact that reputational assessment is an increasingly opaque and secretive process performed by unaccountable firms and technologies introduces some fundamental questions of social justice. David Lyon (2003) defines the hierarchies and divisions that ensue as "social sorting," a process that classifies "people and populations according to varying criteria, to determine who should be targeted for special treatment, suspicion, eligibility, inclusion, access and so on" and leads to a process of assigning a subject's worth and risk in a manner that may have powerful effects on her life chances (20). The sorting is not necessarily meritocratic.

While the facts that contribute to the formation of quantified reputations are seemingly value-free – collected from location histories, retail transactions, e-book reading habits, traffic cameras, government records, etc. – the methods and locations of surveillance that ultimately shape those reputations are implicated by who is motivated to surveil, where they choose to do it, and what or who they choose to observe. Body-worn camera video is a prime example. The choices and practices that create body-worn camera video and the geographic locations where the devices are most likely to be used are tainted by legacies of discrimination, both on the part of police forces that use the devices and the social systems that contribute to social and economic marginalization. Meanwhile, the effect of being captured on a body-worn camera results in an instant unassailable categorization of "police data subject," a typically negative label. In the hands of data brokers, conclusions derived from a particular artifact can echo and amplify whatever bias is attached to the artifact. It follows that connecting body-worn camera video to quantified reputation formation is poised to entrench the existing social sort of marginalization, rendering public spaces as forums of opportunity risk for people struggling to emerge from legacies of dominance and oppression.

Key recommendation: regulate

The continued expansion of public and private surveillance and the inclusion of surveillance data in reputational assessment is a major shift in modern social life. Because this shift has the potential to curtail human autonomy, freedom, and overall well-being, the methods and effects of quantified reputation should invite increased scrutiny by regulators. In the United States at least one form of reputational determination is currently regulated. Numeric credit scores are assigned to U.S. consumers by the three main credit reporting agencies and are used as gatekeeping devices for various types of credit access, as well as various non-credit purposes, including tenancy and employment (Poon 2007; Rice and Swesnik 2013). Until the credit system was subject to federal regulation, it was secretive and reflected discrimination and bias, variously penalizing women, non-Whites, and the non-heteronormative, among others (Dixon and Gellman 2014; Pasquale 2015). Regulation of this form of quantified reputation remains imperfect, but conditions and options for redress for those judged by credit scores are markedly improved over those that existed prior to regulation.

Consumer scores and categories resemble credit scores in their use as quantified gatekeeping devices, but there is no regulatory regime in the United States for them, and that is cause for concern. The companies that market reputation products are generally accountable to no one. They can use any information they choose and are not required to provide consumers with any insight into the formation of their assessments nor any mechanism to correct errors. Reputation products are completely opaque; their data sources and proprietary algorithms shielded by trade secret protections, making it virtually impossible to evaluate the appropriateness or fairness of the data they use and the predictions they offer. The congressional committee that investigated the data broker/reputation industry was rebuffed in their attempts to learn details of how the systems work and what sources of data are used. Other than acknowledging that reputation products rely heavily on information drawn from government records, companies generally refuse to provide details about their methods, citing confidentiality and trade secret protection (Bachman 2013).

Many commentators who have weighed in on the question of state surveillance and data mining often repeat some variation on the adage "if it can be mined, it will be" (Gagné 2016; Nissenbaum 1998; Fertik and Thompson 2015). While a link between body-worn camera video and data mining is only speculation at this point, the possibility that police body-worn camera video could ever be incorporated into the data used by the reputation industry and its consumers raises important questions about surveillance in public spaces and supports a case for regulation. The flow of data produced by police activity into consumer profiles is unlikely to be positive for the subjects of that data. A shadow of suspicion falls on anyone who in any way is included in police data, whether or not they are suspected of any wrongdoing or are convicted of a crime. Profiling algorithms, no matter how sophisticated, are unlikely to be able to differentiate between bona fide criminal suspects captured on body-worn camera video and anyone else made collaterally visible by virtue of their presence in a public space. The data flows of adverse detection have the potential to penalize an already marginalized segment of society by negatively influencing quantified reputational assessments. With the increasing use of quantified reputation in employment, housing, financial, and other important decisions – decisions that can change the course of one's life – the issue of adverse detection implicates the reputation economy as a construct that rewards those who can most easily operate within its norms, and generally obstructs the path to opportunity for those who cannot.

Additional recommendations

Besides general regulation of the reputation economy, there are at least two other recommendations that may minimize the deleterious impacts of police generated data, including body-worn camera video. First, U.S. states must consider the unintended consequences of their public disclosure laws and seek a balance between government transparency and holding data brokers accountable for exploitive information uses that violate the rights of citizens or severely impede their interests. Latanya Sweeney's work on data reidentification (2015) demonstrates the danger to data subjects, including threats to their legitimate confidentiality interests, posed by freely shared, supposedly anonymized, public records. Yet, many state governments are heedlessly sharing ever more data sets without carefully balancing the value of open government with its potentially negative ramifications for the people described by that data.

Many states have taken legislative action to exempt some or all body-worn camera video from public disclosure laws. However, such standards are neither uniform nor absolute. Court challenges to existing access restrictions should be expected and open government/free speech advocates may well prevail. Even so, I suggest that policy makers and courts continue to take the privacy and liberty interests of data subjects into account when deciding just how transparent government records need to be in order to fulfill the original spirit of public disclosure laws without enabling commercial predation.

A second recommendation is to reevaluate U.S. trade secrets protections. This is particularly important for commercial data products used to make life-altering decisions in both the private and public spheres. A commercially attractive feature of quantified reputation is its seamless fit into algorithmic decision-making systems. However, a body of evidence calling into question the ethics of algorithmic decision making is starting to emerge. This includes investigative reporting by Dewey (2016) about an algorithmic tenant screening product that makes tenancy recommendations based on the applicant's social media activity, and an analysis of courtroom software for pre-sentencing risk assessments investigated by Angwin et al. (2016). Both accounts raise doubts about the fairness and accuracy of automated assessments, particularly when used in making high stakes decisions about people's lives. This and other research into algorithmic determination highlights serious questions about digital civil rights and the accountability of companies operating within the reputation economy. The increasing use of reputation products to make crucial decisions requires us to hold data brokers to high standards in order to ensure that long-sought, hard-fought anti-discrimination policies are not wiped away by algorithmic profiling. This will not happen without intervention. A culture of transparency and accountability must be constructed in order to ensure that data subjects can confront the information being used to judge them. The industry will claim that revealing their methods will harm competitiveness and innovation, and there may indeed be short-term costs to bear. But the default of privileging trade secret protections at the expense of social progress and the right of individuals to assert some control over their lives is reckless and unacceptable.

Conclusion

The use of body-worn cameras by the police raises serious and difficult questions for all of us, as exemplified by the hypothetical case of Raina McDonald discussed at the outset of this chapter. On the one hand, calls for greater police accountability are powerful and urgent, and they strike to the heart of long-standing questions about racial equity and human dignity. In recent years, as police killings of unarmed African Americans have been captured on video, the contentious issues of discrimination and excessive force by the police have gained overdue mainstream attention. The possibility that body-worn cameras can address these issues is intoxicating for anyone who has longed for justice and a more equitable culture. On the other hand, body-worn cameras are invasive surveillance devices that cast a panoptic eye on our communities, adversely detecting marginalized populations disproportionately and to their peril. With every addition to the surveillant assemblage, we relinquish a little more of our public lives to surveillance, and along with that, more of our freedoms to our observers.

The transformation of public space into increasingly observed and mediated space follows from the idea that the right to privacy in public is an unreasonable expectation, particularly where surveillance is seen as means of fulfilling moral duties to protect and care for others (Himma 2015; Gooas et al. 2015). Yet, as Michael Froomkin (2000) argued nearly two decades ago, the U.S. legal framework does not make it easy to retain control over sensitive information once it has been collected, which argues for approaching the decision to collect it with caution and presumptive restraint (1463). Joining the necessarily invasive nature of police work with blanket surveillance – that may or may not be effective at making positive changes to police–public dynamics – should cause us to reexamine our failure to recognize the value of privacy in public and the potential risks to human well-being of adverse detection.

I am not suggesting that there is only a single stark choice here. Raina McDonald should not have to choose between having greater insight into how the police in her community are treating her neighbors and risking her quantified reputation by stepping outside her home. Yet we must confront the possibility that the body-worn camera may not fully deliver on its promises and we will have enabled an expansion in surveillance of public spaces for little gain. The distressing fact is that video footage displaying apparently unjustified uses of force by the police has yet to result in major changes to police culture and practices. Perhaps the increased use of body-worn cameras will complete the picture (literally) and will be the change that holds police officers truly accountable for their actions, delivering overdue justice to aggrieved communities. While we await those results, we should be wary of this new addition to the surveillant assemblage, the potential for adverse detection, and the algorithmic enablement of social sorting and discrimination.

Notes

1 Revised Code of Washington (RCW) § 42.56.050.
2 RCW § 42.56.240(14)(a).
3 By "virtually everyone" I mean to refer primarily to people residing in places where internet use, digital communication, and electronic commerce are common features of everyday life.

References

Allen, Anita L. 2012. "Privacy Law: Positive Theory and Normative Practice." *Harvard Law Review Forum* 126: 241.

Angwin, Julia, Jeff Larson, Surya Mattu, and Lauren Kirchner. 2016. "Machine Bias: There's Software Used across the Country to Predict Future Criminals. And it's Biased against Blacks." *ProPublica*, May 23. www.propublica.org/article/machine-bias-risk-assessments-in-criminal-sentencing.

Ariel, Barak, William A. Farrar, and Alex Sutherland. 2014. "The Effect of Police Body-Worn Cameras on Use of Force and Citizens' Complaints against the Police: A Randomized Controlled Trial." *Journal of Quantitative Criminology* 31(3): 509–535. doi:10.1007/s10940-014-9236-3.

Ariel, Barak, Alex Sutherland, Darren Henstock, Josh Young, Paul Drover, Jayne Sykes, Simon Megicks, and Ryan Henderson. 2016. "Wearing Body Cameras Increases Assaults against Officers and Does Not Reduce Police Use of Force: Results from a Global Multi-Site Experiment." *European Journal of Criminology*, May 14. doi:10.1177/1477370816643734.

Bachman, Katy. 2013. "Senate Commerce Report Says Data Brokers 'Operate Behind a Veil of Secrecy' [Updated]." *AdWeek*, December 18. www.adweek.com/news/technology/senate-commerce-report-says-data-brokers-operate-behind-veil-secrecy-154579.

Barbaro, Michael, and Tom Zeller. 2006. "A Face Is Exposed for AOL Searcher No. 4417749." *The New York Times*, August 9. http://query.nytimes.com/gst/abstract.html?res=9E0CE3DD1F3FF93AA3575BC0A9609C8B63.

BBC News. 2015. "Freddie Gray Death: Protesters Highlight Other Police Deaths." April 28, sec. US & Canada. www.bbc.com/news/world-us-canada-30341927.

Blanchette, Jean-François, and Deborah G. Johnson. 2002. "Data Retention and the Panoptic Society: The Social Benefits of Forgetfulness." *The Information Society* 18(1): 33–45. doi:10.1080/01972240252818216.

Boutin, Paul. 2016. "The Secretive World of Selling Data About You." *Newsweek*, May 30. www.newsweek.com/secretive-world-selling-data-about-you-464789.

Cahill, Michael T. 2007. "Retributive Justice in the Real World." *Washington University Law Review* 85: 815.

Chen, Jinyan. 2014. "Gait Correlation Analysis Based Human Identification." *The Scientific World Journal* 2014. doi:10.1155/2014/168275.

"Considering Police Body Cameras." 2015. *Harvard Law Review* 128(6): 1794+.

Council on Crime and Justice. 2007. "Justice, Where Art Though? A Framework for the Future." Minnesota, MN: Council on Crime and Justice.

Dewey, Caitlin. 2016. "Creepy Startup Will Help Landlords, Employers and Online Dates Strip-Mine Intimate Data from Your Facebook Page." *The Washington Post*, June 9. https://www.washingtonpost.com/news/the-intersect/wp/2016/06/09/creepy-startup-will-help-landlords-employers-and-online-dates-strip-mine-intimate-data-from-your-facebook-page/.

Dixon, Pam, and Robert Gellman. 2014. "The Scoring of America: Secret Consumer Scores Threaten Your Privacy and Your Future." World Privacy Forum.

Doyle, Tony, and Judy Veranas. 2014. "Public Anonymity and the Connected World." *Ethics and Information Technology* 16(3): 207–218. doi:10.1007/s10676-014-9346-5.

Fellner, Jamie. 2009. *Decades of Disparity: Drug Arrests and Race in the United States*. New York: Human Rights Watch.

Ferguson, Bob. 2014. "Video And Audio Recording of Communications between Citizens and Law Enforcement Officers Using Body Cameras Attached to Police Uniforms." Washington State. Office of the Attorney General. November 24. www.atg.wa.gov/ago-opinions/video-and-audio-recording-communications-between-citizens-and-law-enforcement-officers.

Fertik, Michael, and David C. Thompson. 2015. *The Reputation Economy: How to Optimize Your Digital Footprint in a World Where Your Reputation Is Your Most Valuable Asset*. First edition. New York: Crown Business.

Fleming, Julia. 2004. "The Right to Reputation and the Preferential Option for the Poor." *Journal of the Society of Christian Ethics* 24(1): 73–87.

Florida Star v. B. J. F. 1989. 491 US 524. US Supreme Court.

Foucault, Michel. 1977. *Discipline and Punish: The Birth of the Prison*. New York: Pantheon Books.

Froomkin, A. Michael. 2000. "The Death of Privacy?" *Stanford Law Review* 52(5): 1461–1543. doi:10.2307/1229519.

Gagné, Marc. 2016. "Does Privacy Still Exist? Data and Privacy Insights with Marc Gagne." *Irish Tech News*. December 19. https://irishtechnews.ie/does-privacy-still-exist-data-and-privacy-insights-with-marc-gagne/.

Garvie, Clare, Alvaro Bedoya, and Jonathan Frankle. 2016. "The Perpetual Line-Up." Center on Privacy & Technology, Georgetown Law. www.perpetuallineup.org/.

Goos, Kerstin, Michael Friedewald, William Webster, and Charles Leleux. 2015. "The Co-Evolution of Surveillance Technology and Surveillance Practices." In *Surveillance in Europe*, edited by David Wright and Reinhard Kreissl. London, New York: Routledge.

Haggerty, Kevin D., and Richard V. Ericson. 2000. "The Surveillant Assemblage." *The British Journal of Sociology* 51(4): 605–622. doi:10.1080/00071310020015280.

Henderson, Wade, and Karen McGill Lawson. 2011. "The Reality of Racial Profiling." www.civilrights.org/publications/reports/racial-profiling2011/the-reality-of-racial.html.

Himma, Kenneth Einar. 2015. "Why Security Trumps Privacy." In *Privacy, Security, and Accountability: Ethics, Law And, Policy*, edited by Adam D. Moore. Lanham, MD: Rowman & Littlefield International.

Hinds, Kate. 2013. "Some Police Departments Embrace Body Cameras." New York, NY: WNYC. www.wnyc.org/story/312260-police-departments-and-body-cameras/?utm_source=sharedUrl&utm_medium=metatag&utm_campaign=sharedUrl.

Hundtofte v. Encarnacion. 2014. 181 Wn.2d 1. Washington State Supreme Court.

Jennings, Wesley G., Lorie A. Fridell, and Mathew D. Lynch. 2014. "Cops and Cameras: Officer Perceptions of the Use of Body-Worn Cameras in Law Enforcement." *Journal of Criminal Justice* 42(6): 549–556. doi:10.1016/j.jcrimjus.2014.09.008.

Jennings, Wesley G., Mathew D. Lynch, and Lorie A. Fridell. 2015. "Evaluating the Impact of Police Officer Body-Worn Cameras (BWCs) on Response-to-Resistance and Serious External Complaints: Evidence from the Orlando Police Department (OPD) Experience Utilizing a Randomized Controlled Experiment." *Journal of Criminal Justice* 43(6): 480–486. doi:10.1016/j.jcrimjus.2015.10.003.

Joh, Elizabeth E. 2016. "Beyond Surveillance: Data Control and Body Cameras." *Surveillance & Society* 14(1): 133–137.

Lyon, David. 2003. *Surveillance as Social Sorting: Privacy, Risk, and Digital Discrimination.* London, New York: Routledge.

Lyon, David. 2007. *Surveillance Studies: An Overview.* London: Polity.

Maass, Dave, and Aaron Mackey. 2016. "5 Ways Law Enforcement Will Use Tattoo Recognition Technology." *Electronic Frontier Foundation.* June 2. www.eff.org/deeplinks/2016/05/5-ways-law-enforcement-will-use-tattoo-recognition-technology.

Miller, Lindsay, Jessica Toliver, and Police Executive Research Forum. 2014. "Implementing a Body-Worn Camera Program: Recommendations and Lessons Learned." Washington, DC: Office of Community Oriented Policing Services.

Newell, Bryce Clayton. 2017. "Collateral Visibility: A Socio-Legal Study of Police Body-Camera Adoption, Privacy, and Public Disclosure in Washington State." *Indiana Law Journal* 92(4): 1329–1399.

Nissenbaum, Helen. 1998. "Protecting Privacy in an Information Age: The Problem of Privacy in Public." *Law and Philosophy* 17(5/6): 559–596. doi:10.2307/3505189.

Norris, Clive, and Gary Armstrong. 1999. "CCTV and the Social Structuring of Surveillance." *Crime Prevention Studies* 10(157–178): 1.

NY Daily News. 2016. "Philando Castile, Alton Sterling and Other Police Deaths." July 7. www.nydailynews.com/news/national/philando-castile-alton-sterling-police-deaths-article-1.2703257.

Pasquale, Frank. 2015. *The Black Box Society: The Secret Algorithms that Control Money and Information.* Cambridge, MA: Harvard University Press.

Pierson, Emma, Camelia Simoiu, Jan Overgoor, Sam Corbett-Davies, Vignesh Ramachandran, Cheryl Phillips, and Sharad Goel. 2017. "A Large-Scale Analysis of Racial Disparities in Police Stops across the United States." *ArXiv:1706.05678* [*Stat*], June. http://arxiv.org/abs/1706.05678.

Poon, Martha. 2007. "Scorecards as Devices for Consumer Credit: The Case of Fair, Isaac & Company Incorporated: Scorecards as Devices for Consumer Credit." *The Sociological Review*55 (September): 284–306. doi:10.1111/j.1467-954X.2007.00740.x.

powell, john a. 2008. "The Color of Opportunity; Race, Place, and Opportunity." *The American Prospect*, October 1, sec. Special Report.

"Predict Crime | Predictive Policing Software." 2016. *PredPol.* www.predpol.com/.

Rice, Lisa, and Deidre Swesnik. 2013. "Discriminatory Effects of Credit Scoring on Communities of Color Symposium: Credit Scoring and Credit Reporting." *Suffolk University Law Review* 46: 935–966.

Rockefeller, Jay. 2013. "A Review of the Data Broker Industry: Collection, Use, and Sale of Consumer Data for Marketing Purposes." Committee on Commerce,

Science, and Transportation: United States Senate. www.commerce.senate.gov/public/?a=Files.Serve&File_id=BD5DAD8B-A9E8-4FE9-A2A7-B17F4798EE5A.

Schmidt, Eric. 2011. "Google Exec Stokes Facial Recognition Privacy Debate." *Biometric Technology Today* 2011(6): 1. doi:10.1016/S0969-4765(11)70101-1.

Singer, Natasha. 2014. "The Scoreboards Where You Can't See Your Score." *The New York Times*, December 27. www.nytimes.com/2014/12/28/technology/the-scoreboards-where-you-cant-see-your-score.html.

Smith, Gavin J. D. 2012. "Surveillance Work(ers)." In *Routledge Handbook of Surveillance Studies*. Hoboken, NJ: Taylor & Francis.

Snyder, Howard N. 2012. "Arrest in the United States, 1990–2010." NCJ 239423. Bureau of Justice Statistics, United States Department of Justice. http://www.bjs.gov/content/pub/pdf/aus9010.pdf.

Solove, Daniel J. 2004. *The Digital Person: Technology and Privacy in the Information Age*. Ex Machina. New York: New York University Press.

Sorrell v. IMS Health Inc. 2011. 131 S. Ct. 2653. Supreme Court.

Stanley, Jay. 2013. "Police Body-Mounted Cameras: With Right Policies in Place, a Win For All." *American Civil Liberties Union*. https://www.aclu.org/police-body-mounted-cameras-right-policies-place-win-all.

Sullivan, Jennifer, and Steve Miletich. 2014. "Costly Public-Records Requests May Threaten SPD Plan for Body Cameras." *The Seattle Times*, November 20. https://www.seattletimes.com/seattle-news/costly-public-records-requests-may-threaten-spd-plan-for-body-cameras/.

Sweeney, Latanya. 2015. "Only You, Your Doctor, and Many Others May Know." *Technology Science*, September. http://techscience.org/a/2015092903/.

Urban Institute. 2016. "Police Body-Worn Cameras: Where Your State Stands." February. http://urbn.is/bodycam.

Weise, Karen. 2016. "Will a Camera on Every Cop Make Everyone Safer? Taser Thinks So." *Bloomberg Businessweek*, July 12. http://www.bloomberg.com/news/articles/2016-07-12/will-a-camera-on-every-cop-make-everyone-safer-taser-thinks-so.

White, Michael D. 2014. "Police Officer Body-Worn Cameras: Assessing the Evidence." E0414256643. Washington, D.C.: Office of Justice Programs, United States Department of Justice. http://permanent.access.gpo.gov/gpo56953/Police%20Officer%20Body-Worn%20Cameras.pdf.

Williams, Timothy. 2015. "Facial Recognition Software Moves From Overseas Wars to Local Police." *The New York Times*, August 12. http://www.nytimes.com/2015/08/13/us/facial-recognition-software-moves-from-overseas-wars-to-local-police.html.

Williams, Timothy, James Thomas, Samuel Jacoby, and Damien Cave. 2016. "Police Body Cameras: What Do You See?" *The New York Times*, April 1. http://www.nytimes.com/interactive/2016/04/01/us/police-bodycam-video.html.

Chapter 7

"The end of privacy as we know it"

Reconsidering public space in the age of Google Glass

Olya Kudina and Melis Baş

Introduction

The introduction of Google Glass in 2013 attracted much public attention and initiated a race for the commercial development of mixed reality goggles. Google Glass (hereafter Glass) is a head-mounted display in the shape of eyeglasses with camera that can overlay and augment the physical world with virtual information. The device is always present in the user's field of sight and demands close attention during each interaction. Glass enables online search, personalized suggestions and navigation in real-time. Also, Glass can take pictures and record the surroundings, share and store them online, while not providing clear signals to the outside. While Glass does not allow a continuous recording, "those who wish to can record, rewind and rewatch more of what they see more easily – and where everyone else can end up recorded as part of the process" (*The Economist*, 2013).

In view of its surreptitious recording capabilities, Glass generated various concerns, primarily related to privacy. Soon after its introduction, "Glass-free zones" emerged on some business premises, where owners considered the video camera embedded in Glass to be a violation of privacy of their clients. Other creative appropriation of Glass included sabotaging the functioning of device by blocking its Wi-Fi connectivity. Even though Google emphasized the privacy-conscious character of Glass as a device providing control over user's information, privacy appeared as an unattained cornerstone for the societal acceptability of Glass.

Google withdrew Glass for redesign in 2015, only two years after its introduction. However, this does not mark the end of privacy concerns related to mixed reality devices. An updated version of Glass appeared in July 2017, currently available only for enterprise use (Levy 2017). Moreover, camera-equipped devices, similar to Glass, increasingly enter the market of mixed reality glasses (e. g. HoloLens by Microsoft (2015), Spectacles by Snap (2017) and EyeTrack Insight by Olympus (2017)). To this end, the debate around Google Glass and privacy remains relevant. In anticipation of the further integration of mixed reality devices into the daily lives of people, it is important to understand the privacy concerns they generate, their foundation and implications. Glass, just as other mixed reality goggles, can be used in different social scenarios, ranging from face-to-face intimate encounters to the use in public space. Conceived as a locus for socialization, interaction, and identity representation, public space embeds both individual, interpersonal and larger group concerns. Therefore, in this chapter, we seek to examine the impact of mixed reality glasses on the nature of public space, and suggest to turn to Glass as a device with history in this regard.

Philosophically, we rely on the theory of technological mediation (Verbeek 2005, 2011) and the thought of Hannah Arendt (1990, 2013) to understand Google Glass better. The theory of technological mediation understands technologies to be in dynamic, mediating relations with the people and the world (Ihde 1990; Verbeek 2005). As such, people and technologies are not independent of each other, because, on one hand, people design technologies with certain intentions; but on the other hand, these same technologies help to shape the perceptions and interpretation of the world and consequently influence the way people act. The ethical consequence of this is that human values are also not independent of technologies – technologies mediate morality (Verbeek 2011). In the current study, we want to understand how privacy as a value takes shape in relation to Google Glass in public space.

We also suggest viewing Glass through the metaphorical concept of a mask, as interpreted by the political philosopher Hannah Arendt. According to Arendt (1990, 2013), a mask concerns the intricate relationship between the identity construction and the appearance of a person in public, allowing a person to appear embodying a certain identity, a crafted position. For Arendt, maintaining a multiplicity of identity representations is an essential part of a public space. Applying the Arendtian concept of a mask to Glass allows us to extend its originally intended two-fold meaning, related to identity construction and representation, by including its function as a mask collector and narrator.

Contrary to Arendt, Glass seems to obscure the plurality of representations of the self in public by defining them through the prism of privacy as control of information. However, the practices Google Glass enables present a rich background for reasoning about privacy, where the logic of control may be one of many possibilities. To understand how people reason about privacy in relation to Glass, we will follow this technology in the practices in enables. We will do so by examining how privacy discussions about Google Glass emerge online, particularly on YouTube. As a space to capture a multitude of public perspectives in the form of videos and comments, YouTube is a valuable platform to understand the way people perceive a technology such as Google Glass, a technology that is not fully present, having been available only to some people and currently available only for enterprise use. Our empirical study will remain explorative because of the selective and constructed character of the comment base we analyze. It nonetheless provides a deeper understanding of how people try to make sense of augmented and mixed reality technologies by presenting a snapshot of privacy discussions on YouTube in relation to Glass.

The structure of our study is as follows. First, we elaborate on the theoretical cornerstone of our study, the theory of technological mediation. Then, we draw on the Arendtian concept of a mask to approach Glass as a technology in the public space and the uncertainty it entails.

Afterwards, we explain the explorative empirical philosophy, with attention to methodology, which we used to study the case of Glass and privacy. Finally, we proceed to the empirical study, where we identify and analyze several practices, depicting how people use or anticipate using Glass; investigate the practice-specific privacy puzzles and present privacy as a porous, contingent and multi-layered value. In the concluding part, we argue for an increased responsibility of Glass users to be aware of and preserve the multiple dimensions of the value of privacy, and thus safeguard the presence of plurality in the public space.

Highlighting the technological in privacy: the theory of technological mediation

Privacy, as a value, frequently features in public debate and policy-making. Despite the dominant legal and corporate discourse on privacy as information control and management (Data Protection Directive 1995), there is no one agreed definition. When we turn to historical accounts of privacy, we see that maintaining privacy has gradually changed with the introduction of new technologies. The mass appearance of photographs in newspapers at the end of the nineteenth century made us experience privacy as "the right to be let alone" (Warren and Brandeis 1890), followed in the 1960s by a focus on personal data protection as a reaction to the emergence and spread of digital data (Solove 2002). In recent decades, the constant use of (personal) devices connected to the Internet – devices that increasingly penetrate every sphere of our life – has extrapolated this effect. In the contexts of these technologies, another understanding of privacy emerged and was expressed at the EU level as "the right to be forgotten" (Mayer-Schönberger 2009). What follows from these observations is that, firstly, privacy is a dynamic value, and, secondly, people attribute meaning to privacy in relation to the technologies around them.

The theory of technological mediation (Ihde 1990; Verbeek 2005) suggests that when in use, technologies mediate the perceptions and actions of people in the world. People embody technologies to the point of considering them an inalienable counterpart of self (e.g. a pair of glasses or a blind-man's cane). The world rarely reveals itself to people directly, but rather with or through technologies (e.g. a thermometer or a screen). Similarly, people interact with technologies to get access to the world (e.g. a cash machine or a smartphone), in which technologies increasingly form a silent background for our existence (e.g. air conditioning or WiFi networks). Overall, the technological mediation theory suggests that it is through technologies – broadly construed – that people act and realize themselves in the world.

Moreover, in a world that is increasingly technological, people and technologies co-shape each other in a constant interaction: "What the world 'is' and what subjects 'are,' arises from the interplay between humans and reality" (Verbeek 2008, 13). Such a mediating phenomenon of technologies carries normative connotations: if technologies mediate the experiences and practices of people, then human values that emerge within them are also mediated by these technologies. Verbeek (2011), inspired by Annemarie Mol (2002), considers human values as interactional. According to Mol (2002), human values are necessarily embedded in practices that enable or contradict them. Different practicalities enact different configurations of what a value means. Verbeek (2011) conceptualized the interdependent nature of human values and technologies in the approach to technological mediation of morality highlighting the interactional, dynamic, and co-shaping nature of human values and technologies.

Following the approach to technological mediation of morality, the value of privacy is not divorced from Google Glass and the practices it enables. Moreover, different practices can enable different manifestations of the same value. It seems that to understand what is at stake with privacy and Glass, we need to identify and interpret the ways in which people use Glass and what it means for public space. We will first elaborate on a particular interpretation of public space that we build on, and then provide a brief analysis of Google Glass and the uncertainty in public space it enables.

Glass as performative uncertainty in public

When we talk about the public space, we predominantly take after Arendt, who extensively studied its nature and purpose. Following Arendt (1990, 198–199), public space is not necessarily a designated physical location; instead, any space can turn into a public space. What turns a space public is a political act: for example, when people gather for the public performance of deeds and the speaking of words, and thus manifest their appearance in the public (Arendt 1990). This political function of a public space is also crucial for the process of identity construction and representation: one has to be prepared for constant observation by others. Even though we embrace Arendt's understanding of the public space as constitutive for the process of identity building and representation, we do not want to limit it to the political context and would like to build on its social value. Therefore, we opt for a broader conception of public space as a locus for socialization, interaction, and identity representation.

When we enter a public space, we implicitly consent to the rules governing it. Increasingly, public authorities consider surveillance an intrinsic feature of the urban landscape. If people want to participate in public spaces, they must trust the entities behind surveillance to manage their data in accordance with legislative norms. However, it has been argued that the ubiquitous penetration of digital devices into everyday life challenges the fitness of the consent mechanism to address the dynamic nature of data flows, as well as data collection, storage, and processing (Matzner et al. 2016; Van den Hoven and Vermaas 2007).

In this chapter, we will not focus on the legal aspects of privacy or personal data protection in relation to Glass. Instead, we are more interested in how the uses of Glass can change and redefine the public place, and, in parallel, whether and how privacy as a value plays a role in it.

A preliminary look at Glass reveals the device as an individual wearable technology, promoting simultaneous connection with both physical and digital environments. Google promoted Glass on the assumption that modern technologies somehow suppress people and prevent them from fulfilling their goals; hence, it fell to Google to fix this with the help of Glass. According to its website, "Our vision behind Glass is to put you back in control of your technology," thus emphasizing the values of empowerment, proactiveness, and control that Glass can provide all users, "from moms to mountain climbers" (Wayback Machine 2015).

The main rationale underpinning the security and privacy policies of Glass concerns control of information (Glass 2013). Here the emphasis is on user empowerment because it is the user who makes decisions about the recording and sharing of data. Behind the spotlight is the fact that all recorded data is synchronized and stored in Google Cloud, making it potentially accessible for the owner of Glass from any device at any time. Google did not specify which information about the user it collects, nor the technical details concerning processing the recorded data.

At the same time, Google did not forget about the people around Glass wearers and noted that it had built in "explicit signals" to notify when Glass is recording, such as: illuminating the screen, providing voice commands, and a built-in LED red light. However, these signals were deemed insufficient and unsatisfactory both by people in the proximity of Glass users (Koelle, Kranz, and Möller 2015); and data protection authorities worldwide (Office of the Privacy Commissioner of Canada 2013). Commenting on the Glass feature to record with the wink of an eye, the company recommended Glass wearers to "use [their] best judgment" (Glass 2015) when capturing their surroundings, thereby trusting users to navigate wisely the social etiquette concerning Glass.

Glass introduces an unparalleled element of uncertainty into the public space, which stems from the fact that there are no clear signals to communicate to the public whether the device is recording at any particular moment. While other personal gadgets can record in public, they require a certain chain of procedures, such as: holding the device, positioning it in a way suitable for recording, and touching buttons to activate or stop the recording. Glass does not require any of these noticeable actions: one can activate the recording with the blink of an eye or by issuing a suitable voice command.[1]

Another significant difference of Glass from the existing devices, such as smartphones, is that it is located in the direct field of sight of the user (above the right eye). When in use, Glass requires close attention to the signals (e.g. navigational aural cues or notifications requiring action); interaction (via voice commands or by tapping the touchpad behind the ear); and concentration (visual cues require a close focus on a screen). Such a technological setup was carefully designed "to cater to *microinteractions*, allowing the wearer to utilize technology while not being taken out of the moment" (Firstenberg and Salas 2014, 11). In practice, however, it required Glass wearers to often look at the screen and to verbally interact with the device, all of which complicated their interaction with the people nearby (Honan 2013; Koelle, Kranz and Möller 2015).

Moreover, owing to its design (imitating a pair of ordinary glasses), Glass invites the public to view it as an ordinary object. However, a qualitative study conducted by Koelle, Kranz, and Möller (2015) indicates that, in fact, its ambiguous design makes bystanders insecure and suspicious of the device because it is difficult to deduce what activity the Glass wearer is engaged in. The study also shows that people in the proximity of a Glass wearer by default assumed the device was recording or taking pictures of them (ibid., 366). Therefore, the design factor makes Glass subject to public suspicion, as neither the form nor designated signals can communicate the specific action and/or intent of the wearer.

The Glass users themselves gave the public a reason to be suspicious. In its short existence, Glass fostered a creation of a new term, "Glasshole," condemning a socially inappropriate behavior of Glass users, often related to secretly taking pictures and videos of people. Shortly after its introduction in 2013, multiple Google Glass etiquettes surfaced in media to teach Glass wearers how to properly use the device in a "polite society" (Sintumuang 2013; Mashable 2013). Certain business owners defined their premises a "Glass-free zone," concerned with the privacy of their visitors (Dickey 2014). Oliver (2014), inspired by a slogan "Privacy is the power to selectively reveal oneself to the world" (Hughes 1993), created a program that identifies the entry of a Glass wearer in a public area (e.g., a restaurant) and cuts off the Wi-Fi connectivity of the device. Most of these activities were preoccupied with privacy in public places and perceived the proximity of a Glass user as a threat to privacy, or, as Greenberg (2014) describes it, "Google Glass's ability to turn ordinary humans into invisibly recording surveillance cyborgs."[2]

If we perceive public space as a place for socialization, interaction, and representation of one's identity, then the introduction of uncertainty in the shape of Glass could have profound restructuring effects on the nature of public space. The renowned analysis on the power of uncertainty in public space to which we turn belongs to Foucault (1977). In a popular interpretation of

Bentham's idea of the Panopticon (Bentham 1791), Foucault considered the uncertainty and unequal conditions it creates to enable the power of one-way surveillance. The single direction of surveillance is constituted by the architectural structure of a panoptic prison: the guards are in a centrally located tower, able to observe the prisoners at any time without being seen by the prisoners. The idea behind the Panopticon is that, even though it is unknown whether the guards are watching the prisoners, the prisoners assume they are always visible and thus model their behavior to the expected standards. The power of uncertainty thus internalizes the surveilling eye within the prisoners themselves and subjugates them.

In relation to Glass, however, we argue that another side of the panoptic coin is also relevant: not so much the internalization of the surveilling gaze, but rather the centralization of it. We argue that the centralizing aspect of observation, collection, and narration of Glass is implicitly at the center stage of the privacy debate around it. Glass allows for an all-encompassing interpretation of the accumulated public narratives from a one-person perspective, taking as a given that these public narratives are detailed recordings, permanently stored and easily accessible. Control of information is the only privacy lens that Glass has. However, people around the Glass wearer may expect privacy on different levels, where the logic of control is but one (Patton 2000). The contextual and dynamic nature of privacy enables a variety of practices and narratives to thrive in public space (Hildebrandt 2006). By highlighting the aspect of control of information in privacy, Glass suppresses the expression of other values, embedded in the fluid and multidimensional concept that privacy is. As such, Glass complicates representation of the multiplicity of captured identities and individual stories that call on privacy in a way different from control of information, essentially restructuring the nature of a public space.

To understand why the reduction of multiple positions in public space challenges its nature and structure, we will seek answers from Arendt as an author who extensively investigated the concept and value of public space.

Glass as a mask collector and narrator: an Arendtian perspective

Following Charles Taylor (2003), the Western political tradition is largely imbued with ideas of classical liberalism. One of the guiding principles of liberal democracy concerns the prevalence of the atomistic individual and his interests, goals, and conceptions of a good life. The society that emerges from the liberal tradition is merely a sum of its component members. The will of a person thus impinges on society with individual moral frameworks and life plans, levelling society to the perspective of one. Taylor argues that the "basic principles [of a liberal society] concern how society should respond to and arbitrate the competing demands of individuals" (ibid., 4).

One of the thinkers who opposed such a liberal atomistic conception of society and outlined its dangers was Hannah Arendt. She argued for an alternative societal layout that brought to the fore the aspect of plurality in human life. According to Arendt, "[t]he hope for man in his singularity lay in the fact that not man but men inhabit the earth and form a world between them" (1990, 174). It was of the utmost importance to Arendt to never single out humans as individuals. Genuine human life was, for her, the one shared with others. Only through these others can we perceive the world in the entirety of its many perspectives.

Plurality is thus the basis for society. An individual's values and motivations take shape through his or her interaction with others in the public space. If a prevalence is attributed to one individual perspective or if plurality of representations is downgraded, society may lose its versatility and end in a totalitarian state (Arendt 1990). Thus, the principles of the democratic society, according to Arendt, should foster the plurality of identities, voices, and perspectives represented in their multiplicity in the public arena.

To facilitate the implementation of the democratic condition for plurality, Arendt turned to the Greek concept of a mask to investigate how unique human identities and voices penetrate the public space. We also turn to the concept of a mask to investigate how varying conceptions of privacy emerged in relation to Glass. Owing to its nature, one can interpret Glass as an additional layer of information consisting of aural and visual cues in the physical environment. Combining this insight with the aesthetic form of Glass, one can think of Glass as a mask that presents the outside environment to the user and the user to the environment in a specific way. In the following sections, we consider the mask metaphor, not so much concerning its fitness for a visual description, but rather pondering its philosophical value in terms of Arendt.

The nature and value of an Arendtian concept of a mask

Arendt used the concept of a mask to distinguish the intricate relationship between the appearance of a persona in public and the construction of identity. For Arendt, the essence of a mask stems from discerning the original meanings of the Latin words *persona* and *personare*. Persona translates to a theatrical mask that an actor wears in tragic plays in ancient Greece; personare adds to this a provision for allowing the actor to speak and be heard. Following Arendt, "The mask as such obviously had two functions: it had to hide, or rather to replace, the actor's own face and countenance, but in a way that would make it possible for the voice to sound through" (Arendt 1990, 106). Thus, wearing a mask enables a person to appear in public embodying a certain identity, a contextually defined role, a crafted position.

According to Arendt, every time we enter the public space, we perform a persona, a public self, constitutive of who we are or who we wish to be seen as. While in the public space, our private self hides behind a mask and we become a political actor. In the eyes of Arendt, humans constitute who they are by entering the public space while acting and speaking in it. In her prominent work, *The Human Condition*, Arendt argues (2013, 50): "For us, appearance – something that is being seen and heard by others as well as by ourselves – constitutes reality." This "appearing in front of others" is the key for genuine identity building. Whereas private space allows us to be true to ourselves – in terms of taking care of our physical needs and contemplating who we are – the public space is the only place where we can act on that "who", that mask, that persona we want to establish. In other words, entering a public space is the only way we can shape who we are in all its multiplicity.

It is important to emphasize that, although we carefully choose the mask to represent us in public, it is for others to interpret. The act of appearance in public becomes meaningful only when others acknowledge and narrate it, relying on what they previously saw or heard. Moreover, according to Arendt, the mask is not only a product of individual thinking; rather, it is also co-shaped by "the exigencies of the play" that the mask has to address. Considering the creation of a legal personality, Arendt said:

> The distinction between a private individual in Rome and a Roman citizen was that the latter had a persona [...]; it was as though the law had affixed to him the part he was expected to play on the public scene, with the provision, however, that his own voice would be able to sound through.
>
> (Arendt 1990, 107)

Thus, public space serves as a stage that gives durability to human life and the world people create through actions. People appear in front of others to be recognized. Acting in public – by becoming visible to others – reveals and exposes the identity of a subject. This disclosure *makes* the subject: becoming visible to others through action not only means revealing, but also performing, an identity. Public space is thus essential for the construction and performance of an identity.

For Arendt, we can remain in our private domain and contemplate which mask we will wear before entering the public space, but we can never control what others perceive of us. She continues:

> It [the mask] can be hidden only in complete silence and perfect passivity, but its disclosure can almost never be achieved as a wilful purpose, as though one possessed and could dispose of this "who" in the same manner he has and can dispose of his qualities. On the contrary, it is more than likely that the "who", which appears so clearly and unmistakably to others, remains hidden from the person himself.
>
> (Arendt 2013, 179)

It is because of this revealing one's "who"-ness in front of others that we can never purposefully control who we are. It will always be left to the discretion of someone else, who "in lieu of acting or appearing in his or her own right, watches, remembers and later narrates" (Barbour and Zolkos 2011, 1).

The mask thus shapes our appearance and identity. Although the public persona is, as such, never at our disposal, it is also not at the disposal of any other individual. The plurality of masks in the public space is the safeguard against the monopolizing power to define an individual. It is precisely this opportunity for representation in public that a mask provides, which also helps us to enable and safeguard a democratic condition of plurality.

Applying the concept of a mask to Google Glass

If we were to interpret Glass as the previously described two-fold mask, this would break down into two mutually informing components. The first component would be the image Google wished to project on the Glass user, which would have at least a partial performative power on the way the outside public would interpret the wearer of Google Glass. The second component would be the capacity for self-expression that the user gained by using this device. We will examine these constitutive elements of a persona in the case of Glass.

Looking at Glass through the prism of a mask helps us discern its two-fold function as providing an ability to construct and express distinct identities. However, taking into account the control-based rationale Google ascribed to the persona of a Glass user, combined with the recording functions of the device, allows us to extend the original two-fold concept of a mask to a third dimension. Glass allows its users not only to present a certain mask in public, but also to collect the masks of those around them. Moreover, the opportunity to store the detailed captured information in Google Cloud and to retrieve it at anytime from anywhere gives the Glass user an unprecedented ability to define collective and individual narratives from a single-person perspective, thus controlling the information of other people.

The control-based mode of thinking that Google envisions for the users of its services mirrors an atomistic understanding of society, in which the individual's logic prevails. The control model, as promoted by Google for its reasoning about the privacy of the Glass user, assumes that privacy is a domain pertaining to an individual, the domain that the user can control if desired. It is this model of reasoning about privacy that Glass invites its users to bring into the public space. Moreover, the logic of control is also performative in how Glass users consider themselves, those around them, and the information captured by the device about such individuals. The third dimension of a mask – that of a mask collector and narrator – could thus seemingly challenge the plurality of representation as one of the pillars of identity construction and democratic society that a mask should help constitute in the first place.

In the following section, we thus want to examine the general applicability of the privacy-as-control-of-information model in the public space of which Google Glass is a part. In this, we were inspired by the technological mediation theory and a mix of methodologies to explore Glass as a technology on the brink of innovation.

How to study the technological mediation of privacy: a note on methodology

In this chapter, we want to understand the privacy implications of Glass for the public space and to understand better the way people make sense of privacy in relation to this technology. Following the technologically mediated, interactional take on values of the mediation theory, values manifest themselves in human practices, real or envisioned. In order to understand the various dimensions of privacy, mediated by Google Glass, we intend to follow Glass in the practices it enables. However, Glass is a "borderline technology", a technology at the brink of introduction: although its earlier versions already entered the market and some people experienced Glass first-hand, the device is currently under redesign for general use and the final version is subject to anticipation. The question then is how to trace and examine the practices of a technology that is not fully present. This section presents a description of the explorative empirical philosophy, with attention to methodology, which we used to study the case of Google Glass and privacy.

We suggest the study of interaction in online communities as one way to study practices with emerging technologies, such as Glass. Even though the first version of Glass was available for a limited period of time (2013–2015), it was widely discussed online. In this chapter, we will examine the textual interaction – in comments – of an online community on YouTube in reaction to a video depicting a social etiquette for Google Glass. YouTube, as a social network website based on user-generated video content, invites an open discussion not only of the content but also of any topic provoked by the video that is relevant to the commenter (Chenail 2011). One of the limitations of the YouTube comment study is that videos are, in a sense, staged interactions to which commenters react. However, free choice of language style and expression in comments under videos allow participants to engage with what they have seen on their own terms (Veen et al. 2011), which presents great value for this study.

We shall analyze a Google Glass etiquette interpretation of Mashable, an online technology-review platform. In a 1 min. 46 sec. YouTube video, the company depicted in a satirical way why some Glass users are referred to as "Glassholes and how to avoid being one" (Mashable 2013). A series of provocative scenarios depict socially inappropriate behaviours with Glass, such as secretly taking videos and pictures on a date or in a toilet, cheating in games by consulting the search engine, etc. The etiquette guide of Mashable engages the viewers, reviewing certain Glass-fostered practices, and asks for feedback, thus presenting an interest for the research. It was also the first video-etiquette concerning Glass to generate considerable attention, collecting 1,415,003 views and 2,064 comments (as of June 2014)[3], all processed for this work.

The data collection and analysis proceeded manually. A close reading of the comments allowed us to select the final empirical base, consisting of 96 comments, which describe or react to a certain practice with Google Glass. The open nature of YouTube did not require registration to access the comments, all of which were publicly available. To ensure the respect for the autonomy and dignity of the commenters, we anonymized their names and aliases, as well as removed any identifying information, such as date, time and location, from which the comment was made (Markham and Buchanan 2012; Hewson and Buchanan 2013). We preserve the original spelling.

In the analysis of the data, we used coding and thematic analysis to approach the data systematically. We also used elements of Discursive Psychology and Conversational Analysis (DP and CA) method (Potter 1996; Edwards and Potter 1992) to understand in which context the commenters invoked privacy. With this mix of methodology in mind and inspired by the theory of technological mediation, we approach the way people appropriate Glass in online conversations on YouTube.

Approaching the privacy puzzle of Google Glass: an empirical investigation

The prior conceptions of privacy that we carry into our investigation include both the legal and corporate – privacy as a right, privacy as control of information (see, e.g., Glass 2013; Data Protection Directive 1995; Solove 2002); and the social understanding of privacy as a value, dedicated to enabling identity formation and representation, as well as interaction with others (see, e.g., Regan 2002; Steeves 2009; Roessler and Mokrosinska 2013). In combination of the two, we see an intrinsic importance of privacy for public space. We investigate a wide range of practices with Google Glass that could manifest the privacy attributes mentioned above. We thus invite a reading of the excerpts from YouTube comments below in the context of privacy, even though the commenters might not discuss privacy explicitly.

To examine how privacy (as a theme) springs up in the YouTube comments, we will present the findings, organized thematically, to show how privacy discussions differ according to context. Our analysis identified a rich and complex narrative about privacy as a value. Particularly, the commenters discussed privacy as a limited access to the self ("Addressing the GlassHole onslaught"); privacy of personhood; privacy of communication; privacy as civil inattention ("You should be on guard!"); privacy in relation to experience and memories; identity building; activity; and control of information ("The end of privacy as we know it"). Below, for matters of space limitation, we present three of these privacy conceptions, accompanied by an analysis guided by technological mediation theory. We will examine the nature of practice the commenters describe or the tension arising from the use of Glass, followed by an inquiry into whether and why privacy is important for the particular practice. Finally, at the end of each section, we will discuss how Glass as a technology relates to the plurality of identity representations and perspectives on privacy in a public space.

We suggest treating the comments below as exemplifying instances of people enacting multiple understandings of privacy when considering Google Glass. They provide a snapshot of privacy discussions on YouTube in relation to Glass, indicating certain trends in privacy formulations "at one place in time" (Potts 2014). As such, we invite the reader to treat the following analysis not as generalizable but explorative, highlighting the aspects of privacy that are important to people in relation to Glass and how people try to make sense of new technologies such as Glass.

Privacy as control of information: "The end of privacy as we know it"

Excerpt 1

> **Commenter 1**: Am I the only one who thinks anyone wearing these might as well be a government cyborg? these are going to destroy any sense of privacy, send it directly to google (aka the government.)
>
> **Commenter 2**: You must be stupid to buy this. Putting your whole life and privacy in the hands of a personal data-hungry company like Google.
>
> **Commenter 3**: Get used to it, Facebook, and even YouTube has your private information (Google is YouTube). If you're really that paranoid then don't do a half job, abandon the internet completely.
>
> **Commenter 4**: These will end up being abused by the police and government so damn much, the end of privacy as we know it. Plus everything you do and say will be recorded in public places now, its scary to even think about.
>
> **Commenter 5:** I'm sorry, those who pull these kinds of stunts would more than likely get their snotbox busted by someone who isnt cool with it. Google glass with caution I'm just sayin'.

The excerpt above depicts the conflict between wanting to control the information about the self and seemingly not being able to do so because of the ambiguous role of the company behind Glass. Margulis (2003) referred to this as a loss of psychological control over privacy and warned of its detrimental effects to mental well-being. The commenters suggest that as soon as people start using digital devices with Internet connectivity, they should accept losing control over their privacy and submitting their lives to the control of corporations. The obscured corporate practices of data collection and management, along with the uncertainty over the use of Glass by government entities and private individuals, beget an environment of distrust.

This complicates the control of information for people around Glass wearers because they cannot determine what information is shared and with whom, and cannot ensure that only designated pieces of information are transferred to other parties. The ambition for privacy control closely relates to the values of autonomous construction of the self (Hildebrandt 2006). However, the situations discussed in Excerpt 1 are characterized by the lack of control and the many strategies people develop to maintain it, ranging from limiting the use of Internet services (*Commenter 3*) to acts of physical aggression towards Glass users (*Commenter 5*).

In this instance, both the technology (Glass) and the company behind it (Google) mediate the way the commenters reason about privacy. In the eyes of the commenters above, by choosing to wear Glass, its user takes along the baggage of public perception and distrust of the corporate practices behind Glass. This positions the Glass wearer as an adversary in the eyes of a wider public, capable of active measures to sustain the idea of control over their information. In Excerpt 1, we witness an understanding of privacy that most closely relates to the suggested rationale of Google and accords with the corporate and legal definitions: privacy as control of information. At the same time, there is a significant difference: whereas the corporate and legal bodies position the understanding of privacy as control of information proportionate to the current state of affairs, people around Glass users perceive it as a desirable, but currently an unattainable, safeguard.

Privacy as civil inattention: "You should be on guard"

Excerpt 2

> **Commenter 6**: Lack of privacy comes in many flavors … now we have people who can take your picture while non-surreptitiously (you should be on ... guard when addressing someone you don't know who is wearing Glass) facing you.
>
> **Commenter 7**: What the not so distant future holds, more know it all perverts …?

Commenter 6 and 7 both talked about how the nature of public space changes with the introduction of Glass, owing to the suspicion it brings and the inevitable hampering effects on interaction and sociality in the public space. Goffman (1978) conducted a seminal sociological analysis of interaction and communication, and, just as Arendt, relied on the concept of a mask. He drew our attention to the vital importance for the socialization process of the ability to switch between different social roles, "the expressive mask[s] … in face-to-face interactions" (ibid., 73), relative to varying contexts.

Roessler and Mokrosinska (2013) drew on Goffman's work and investigated the social dimension of privacy with an emphasis on sociality. They determined that attending to the norm of privacy as civil inattention enables the process of switching between social masks, which underpins sociality in public. They refer to the work of Nagel to define privacy as civil inattention as "when one's own private affairs are kept private, as well as when respect and reserve are displayed towards others" (Roessler and Mokrosinska 2013, 782). The authors ultimately suggest that only by displaying indifference and civil disregard of strangers can one maintain an expectation of privacy in public.

In a similar vein, Tonkiss (2003) presents an ethics of indifference as a necessary condition for coexisting in the public space. Inspired by the ideas of White on "the gift of loneliness and the gift of privacy" (White 2011, 19) in a city, Tonkiss devises the foundation of the ethics of indifference "as an ethical relation between subjects – one premised less on the 'face-to-face' relations of community than on the 'side-by-side' relations of anonymity" (2003, 298). An absence of excessive curiosity from the surrounding people is thus a foundation for an ethics of indifference. Such an ethics of "look[ing] past a face" (ibid., 301) translates into respect for people and essentially enables the multitude of identities to inhabit the public space simultaneously.

Following *Commenter 6*, the ethics of indifference no longer permeates the public space: one cannot blindly trust strangers to regard you with civil inattention in the age of personal gadgets and Internet connectivity. Personal gadgets invite people to peek into the lives of others and even take photos or make video recordings of them. *Commenter* 7, on the other hand, assumes that covert observation and monitoring will become a part of the future everyday life, accompanied by Glass users' frequent engagement with the search engine for purposes of information search and verification. No face, no mask can go unnoticed in these grim anticipations of society with Google Glass, where this technology monitors, inspects and singles out.

The comments in Excerpt 2 thus concern the perceived lack of control over information about oneself, conceived here in the form of pictures or videos that a Glass user can (secretly) take of a person. Mirroring the argument we have made above, the uncertainty over the use of Glass in public and the default assumption that users utilize the device for recording (Koelle, Kranz, and Möller 2015), mediates the understanding of privacy in public. Here the commenters invoke privacy as the one called to protect the valuable practices of switching between the identity masks and interacting without suspicion in the public space.

On one hand, Glass mediates the ethics of indifference in a disabling way. On the other hand, by allowing people around Glass wearers to identify what is important to them, the device enables the meaning of privacy as civil inattention, where the (assumed) excessive curiosity of Glass wearers is not desirable. The idea of privacy as civil inattention that the commenters put forth thus does not fit the conception of privacy as control of information. As such, the mask of suspicion arises on the faces of people around Glass wearers that calls for identifying new ways of safeguarding the ethics of indifference and privacy as civil inattention.

Privacy of activity and identity construction: "[Don't] look over my shoulder"

Excerpt 3

> **Commenter 8**: Yes, I would use Google Glass A LOT, mainly considering the fact that it offers tremendous privacy, whereas when I use my phone or iPod Touch, people tend to get nosy and look over my shoulder to see what I'm doing, and I find that more annoying tha[n] most other things.

Commenter 8 introduced the issue of maintaining privacy in public when using personal gadgets and how the curiosity of bystanders interferes in his/her activities. *Commenter 8* highlighted the importance of maintaining privacy as civil inattention, as manifested by the condemnation of the way bystanders look over our shoulders in public. However, the presentation of this perspective profoundly differs from the findings above: whereas the people around a Glass wearer devise privacy as civil inattention and address Glass as the technology threatening it, actually or potentially, (potential) Glass users turn to the mediating ability of the device to sustain the ethics of indifference.

Commenter 8 viewed Glass as a technical solution to bring the ethics of indifference back, to repair the situation when people around you can furtively observe your behavior and activities. From the user perspective, Glass invites the ethics of indifference by providing a physical barrier to prevent overlooking how a user interacts with it. A bystander would see a person interacting with his or her device, but would be unable to trace the details of such an interaction. *Commenter 8* emphasized the value of being left to engage with the device without the monitoring eyes of strangers.

Closely related to this argument is another dimension of privacy this excerpt discloses: privacy of identity construction, where "privacy [...] protects the individual's interest in becoming, being, and remaining a person" (Reiman 1976, 44). In the comment above, lamenting about people's excessive interest in the affairs of others is the implicit call for the values of identity and self-expression that privacy is called upon to protect. *Commenter 8* is glad that Glass grants a necessary solitude for private activities, to shape varying masks and to perform those in public. The limited access to the self that Glass provides enables privacy to emerge here also concerning identity building. The design of the device provides a user with a necessary space to shape a persona for public appearance without a gaze of a bystander. In this sense, privacy of activity also closely relates to privacy of identity because Glass mediates both the emergence of the masks the user can assume and the specific meaning of privacy that helps enable the emergence of these masks.

Privacy as civil inattention and the ethics of indifference appear differently from the vantage point of Glass users than from the perspective of people around them. Whereas for people around Glass users the device debilitates the realization of civil inattention in public, the same device facilitates the constitution of the mediating norm of indifference where Glass users are concerned. This instance demonstrates the importance of considering multiple perspectives and contexts when discussing the value of privacy.

Toward increased responsibility by Google Glass users

In this chapter we have analyzed the impact of mixed reality technologies, such as Glass, on the nature of the public space. Equipped with the concept of a mask as interpreted by Arendt, we investigated the nature of Glass: first, considering the intended two-fold functionality of a mask – to promote diverse identities while allowing distinct voices to sound through them; and second, by extending the function of Glass as a mask to that of mask collector and narrator. Following Foucault's investigation of the Panopticon, we positioned Glass as a centralizing locus of surveilling gaze. This seemed to shadow the plurality of representations enacted in the multiple contexts enabled by Glass by defining them through the prism of privacy as control of information. Moreover, the introduction of Glass challenged the conception of the public space as a stage for identity construction and realization, for which a plurality of perspectives is a constitutive condition. We identified the promotion of Glass as granting the users control over their information, while generating uncertainty and lack of control for the people around Glass users to be at the core of suppressing the multiplicity of representations in public.

With the help of the theory of technological mediation, we questioned the fitness of a generalizing legal and corporate privacy conception for the various contexts Glass enabled in the public space. The mediation theory helped us to bring back to the fore the plurality of context-dependent privacy representations and privacy-entangled values, such as sociality, identity construction, autonomy, limited access to the self, civil inattention and respect. The identified privacy formulations only rarely fit into the dominant discourse on privacy as control of information. Rather, the identified technological practices indicated a lack of control and demonstrated how Glass users and people around them tried to deal with it.

Relating our findings to the nature of the public space, we argue that the introduction of Glass is bound to mediate the way we interact, socialize, and represent ourselves in public. We have identified several mediating effects of Glass that could play a role in reconfiguring the public space: elements of suspicion, uncertainty, ambiguity, and, consequently, a desire to control information about the self. Despite the immediate considerations as to the possible detrimental effects to identity construction, representation, and communication, we argue that Glass also fosters new ways to define and represent oneself. Utilizing the analytical force of the theory of technological mediation helped us to uncover the multiplicity of the technology-fostered practices, social masks, and formulations of privacy that remain hidden when employing the narrow conception of privacy as control of information. This plurality allowed the users to identify themselves relative to the new contexts enabled by Glass and to discern what is important for them in relation to it. The public space emerges amidst all of the identified negotiations concerning the masks of identity and reasoning about privacy.

Applying the all-encompassing conception of privacy as control of information to the multiplicity of practices and contexts that Google Glass enables in public places would impoverish the nature of public space and monopolize it by looking at it strictly through the mask of control. We argue that the plurality of identities, values, and representations that the theory of technological mediation and the thought of Hannah Arendt helped to uncover, necessarily enriches the public space, fostering new ways and strategies for interaction and appearance in public.

Privacy as a value in relation to Google Glass appears as multiple, porous and contingent upon the practices where it is enacted. We identified that reasoning about privacy in relation to Glass is not limited to the understanding of privacy as control of information, but also includes the conceptions of privacy as civil inattention, privacy of activity and identity construction. If we consider privacy strictly in terms suggested by Google's narrative – as control of information – then Glass is indeed "the end of privacy as we know it." However, the theory of technological mediation has helped us to reveal the plurality of other available narratives that Glass enables. This marks a different starting point in understanding and reasoning about privacy in public space.

Following Arendt, a multiplicity of perspectives and representations is essential for effective functioning of a public place. In our analysis, we showed how a one-sided interpretation of privacy as control of information can suppress other manifestations of privacy, related to the values of civil inattention and identity construction. In correspondence with the spirit of technological mediation, people have to take responsibility for the way technologies mediate the space around them, their values and life in general (Verbeek 2011). Based on the presented discussions of how technologies co-shape human lives, Glass designers and users should assume the "moral responsibility to actively shape th [e] lives in accompaniment with [...] technologies" (Kiran, Oudshoorn, and Verbeek 2015, 5). It is in relation to Glass that particular user practices, contexts, values and identities are shaped and performed. The theory of technological mediation could thus offer one possible answer as to what to do with the intensive flow of emerging virtual, augmented, and mixed reality gadgets in the face of continuing concerns over privacy. Rather than completely disengaging with such devices or regulating them to the point of suffocation, when designing and using these technologies, particular attention has to be paid to the way human-technology ensembles are enacted. As such, both the designers and users of Google Glass must take responsibility for the practices and actions they produce together with this technology.

Acknowledgements

We would like to thank Tobias Matzner, Sven Nyholm and the anonymous reviewers for commenting on the earlier versions of the chapter. Part of the research for this chapter was financed by the Netherlands Organisation for Scientific Research (NWO), as one of us is part of the research program, "Theorizing Technological Mediation: toward an empirical-philosophical theory of technology" with project number 277-20-006.

Notes

1 One can, however, also start the recording by touching the designated buttons or trackpad.
2 It is interesting to note how the behavior of early Glass adopters contradicts the idea of Steve Mann, involved in the development of augmented reality goggles since 1970s. According to Mann (2004), it is precisely the ability for a lifelong continuous capture that presents a value of the device, because this enables "sousveillance," an inverse form of surveillance. As such, this technology is highly democratic and presents a balancing power to government and corporate surveillance: "It is this contextual integrity of the evidence, combined with a personal right and responsibility of individuals to preserve evidence, that sets forth an equilibrium between surveillance and sousveillance" (ibid., 2). People around Glass users, however, perceived Glass as a source of surveillance and attempted to counterbalance it with creative appropriations.
3 In April to June 2014, when we conducted the empirical stage for this study, the number of comments available below the video was 2,064. However, during the review of this study in 2017, the number of comments below the same video significantly decreased, to 524. A possible explanation of this could be a filtering policy of YouTube, where administrators remove the content (video and comments) containing spam, hate speech, etc. (YouTube 2017). A lot of the original 2,064 comments indeed contained spam, which we filtered out manually before selecting comments for analysis. However, the 96 comments taken for a close analysis for this study remain intact as of March 29, 2017.

References

Arendt, Hannah. 1990. *On Revolution*. 2nd ed. London: Penguin Books. Original edition, 1963.

Arendt, Hannah. 2013. *The Human Condition*. 2nd ed. Chicago, IL: University of Chicago Press. Original edition, 1958.

Barbour, Charles, and Magdalena Zolkos. 2011. "Action and Appearance: an Introduction." In *Action and Appearance: Ethics and the Politics of Writing in Hannah Arendt*, edited by A. Yeatman, P. Hansen, M. Zolkos, and C. Barbour, 1–9. New York: Continuum.

Bentham, Jeremy. 1791. *Panopticon or the Inspection House*. Dublin: T. Payne.

Chenail, Ronald J. 2011. "YouTube as a Qualitative Research Asset: Reviewing User Generated Videos as Learning Resources." *The Qualitative Report* 16(1): 229–235.

Data Protection Directive. 1995. "Directive 95/46/EC of the European Parliament and of the Council of 24 October 1995 on the protection of individuals with regard to the processing of personal data and on the free movement of such data. COM/95/ 375 COM/92/422 COM/90/314–2." European Parliament, Council of the European Union.

Dickey, Megan Rose. 2014. "A Restaurant Claims Google 'Glassholes' Hurt Its Reviews." *Business Insider*, May 24. Accessed March 26, 2017. www.businessinsider.com/google-glassholes-bad-reviews-2014-5?international=true&r=US&IR=T.

Edwards, Derek, and Jonathan Potter. 1992. *Discursive Psychology*. Vol. 8: London: Sage.

Firstenberg, Allen, and Jason Salas. 2014. *Designing and Developing for Google Glass*. Sebastopol, CA: O'Reilly Media, Inc.

Foucault, Michel. 1977. *Discipline and Punish: The Birth of the Prison*. Vintage.

Glass. 2013. "Glass Security & Privacy." *Google Glass Website*. Accessed June 3, 2015. https://sites.google.com/site/glasscomms/faqs#GlassSecurity&Privacy.

Glass. 2015. "Wink." *Google Glass Support Website*. Accessed June 3, 2015 from https:// support.google.com/glass/answer/4347178?hl=en.

Goffman, Erving. 1978. *The Presentation of Self in Everyday Life*. Harmondsworth: Penguin Books.

Greenberg, Andy. 2014. "'Cut off Glassoles' Wi-Fi with this Google Glass Detector." *Wired*, June 3. Accessed March 26, 2017. www.wired.com/2014/06/find-and-ban-glassholes-with-this-artists-google-glass-detector/.

Hewson, Claire, and Tom Buchanan, eds. 2013. *Ethics Guidelines for Internet-mediated Research*. Leicester: The British Psychological Society.

Hildebrandt, Mireille. 2006. "Privacy and Identity." In *Privacy and the Criminal Law*, edited by E. Claes, A. Duff, and S. Gutwirth, 61–104. Antwerp and Oxford: Intersentia.

Honan, Mat. 2013. "I, Glasshole: My Year With Google Glass." *Wired*, June 3. Accessed March 26, 2017. www.wired.com/2013/12/glasshole.

Hughes, Eric. 1993. "A Cypherpunk's Manifesto." *Activism: Cypherpunks*, March 9. Accessed March 26, 2017. www.activism.net/cypherpunk/manifesto.html.

Ihde, Don. 1990. *Technology and the Lifeworld: From Garden to Earth*. Indiana University Press.

Levy, Steven. 2017. "Google Glass 2.0 is a Startling Second Act." *Wired*, July 18. Accessed October 25, 2017. www.wired.com/story/google-glass-2-is-here.

Kiran, Asle H., Nelly Oudshoorn, and Peter-Paul Verbeek. 2015. "Beyond Checklists: Toward an Ethical-constructive Technology Assessment." *Journal of Responsible Innovation* 2(1): 5–19. doi:10.1080/23299460.2014.992769.

Koelle, Marion, Matthias Kranz, and Andreas Möller. 2015. *Don't look at me that way!: Understanding User Attitudes Towards Data Glasses Usage*. Proceedings of the 17th International Conference on Human-Computer Interaction with Mobile Devices and Services (362–372). ACM.

Mann, Steve. 2004. "Continuous Lifelong Capture of Personal Experience with EyeTap." Proceedings of the 1st ACM workshop on "Continuous archival and retrieval of personal experiences", October 15, 1–21.

Margulis, Stephen T. 2003. "Privacy as a Social Issue and Behavioral Concept." *Journal of Social Issues* 59(2): 243–261.

Markham, Annette, and Elizabeth Buchanan, eds. 2012. *Ethical Decision-making and Internet Research: Recommendations from the AoIR Ethics Working Committee (Version 2.0)*. Association of Internet Research. Accessed June 30, 2015. http://aoir.org/reports/ ethics2.pdf.

Mashable. 2013. "Google Glass: Don't Be a Glasshole." *Mashable YouTube Channel.* Accessed June 1, 2015. www.youtube.com/watch?v=FlfZ9FNC99k.

Matzner, Tobias, Philipp K. Masur, Carsten Ochs, and Thilo von Pape. 2016. "Do-It-Yourself Data Protection – Empowerment or Burden?" In *Data Protection on the Move*, 277–305. Springer.

Mayer-Schönberger, Viktor. 2009. *Delete: The Virtue of Forgetting in the Digital Age.* Princeton, NJ: Princeton University Press.

Microsoft Corporation. 2015. "HoloLens." *HoloLens Website.* Accessed June 16, 2015. www.microsoft.com/microsoft-hololens/en-us.

Mol, Annemarie. 2002. *The Body Multiple: Ontology in Medical Practice.* Duke University Press.

Office of the Privacy Commissioner of Canada. 2013. "Data Protection Authorities Urge Google to Address Google Glass Concerns." *Office of the Privacy Commissioner of Canada [press release]*. Accessed April 10, 2018. www.priv.gc.ca/en/opc-news/news-and-announcements/2013/nr-c_130618/.

Oliver, Julian. 2014. "Find a Google Glass and Kick It From the Network." *Julian Oliver [personal blog]*, May 30. Accessed March 26, 2017. https://julianoliver.com/output/log_2014-05-30_20-52.

Patton, Jason W. 2000. "Protecting Privacy in Public? Surveillance Technologies and the Value of Public Places." *Ethics and Information Technology* 2(3): 181–187.

Potter, Jonathan. 1996. *Representing Reality: Discourse, Rhetoric and Social Construction.* Sage.

Potts, Amanda. 2014. "'LOVE YOU GUYS (NO HOMO)'." *Critical Discourse Studies* 12(2): 163–186. doi:10.1080/17405904.2014.974635.

Regan, Priscilla M. 2002. "Privacy as a Common Good in the Digital World." *Information, Communication & Society* 5(3): 382–405.

Reiman, Jeffrey H. 1976. "Privacy, Intimacy, and Personhood." *Philosophy and Public Affairs* 6(1): 26–44.

Roessler, Beate, and Dorota Mokrosinska. 2013. "Privacy and Social Interaction." *Philosophy & Social Criticism* 39(8): 771–791.

Sintumuang, Kevin. 2013. "Google Glass: An Etiquette Guide." *The Wall Street Journal*, May 3. Accessed March 26, 2017. www.wsj.com/articles/SB10001424127887323982704578453031054200120.

Solove, Daniel J. 2002. "Conceptualizing Privacy." *California Law Review* 90(4): 1087–1155.

Spectacles. 2017. *Spectacles Website.* Accessed March 30, 2017. www.spectacles.com.

Steeves, V. 2009. "Reclaiming the Social Value of Privacy." In *Privacy, Identity, and Anonymity in a Network World: Lessons from the Identity Trail*, edited by I. Kerr, V. Steeves, and C. Lucock, 191–208. New York: Oxford University Press.

Taylor, Charles. 2003. "Cross-purposes: The Liberal-communitarian Debate." In *Debates in Contemporary Political Philosophy: An Anthology*, edited by D. Matravers and J. Pike. London: Taylor & Francis, Routledge.

The Economist. 2013. "The People's Panopticon." *The Economist [briefing]*, November 14. Accessed April 10, 2018. www.economist.com/news/briefing/21589863-it-getting-ever-easier-record-anything-or-everything-you-see-opens.

Tonkiss, Fran. 2003. "The Ethics of Indifference Community and Solitude in the City." *International Journal of Cultural Studies* 6(3): 297–311.

Van den Hoven, Jeroen, and Pieter E. Vermaas. 2007. "Nano-technology and Privacy: On Continuous Surveillance outside the Panopticon." *Journal of Medicine and Philosophy* 32(3): 283–297.

Veen, Mario, Bart Gremmen, Hedwig te Molder, and Cees van Woerkum. 2011. "Emergent Technologies against the Background of Everyday Life: Discursive Psychology as a Technology Assessment Tool." *Public Understanding of Science* 20(6): 810–825.

Verbeek, Peter-Paul. 2005. *What Things Do: Philosophical Reflections on Technology, Agency, and Design*. Penn State University Press.

Verbeek, Peter-Paul. 2008. "Obstetric Ultrasound and the Technological Mediation of Morality: A Postphenomenological Analysis." *Human Studies* 31(1): 11–26. doi:10.1007/s10746-10007-9079-0.

Verbeek, Peter-Paul. 2011. *Moralizing Technology: Understanding and Designing the Morality of Things*. University of Chicago Press.

Warren, Samuel D., and Louis D. Brandeis. 1890. "The Right to Privacy." *Harvard Law Review*: 193–220.

Wayback Machine. 2015. "Glass. What it Does." *The Internet Archive*. Accessed March 26, 2017. https://web.archive.org/web/20150115191209/www.google.com/glass/start/what-it-does/.

White, Elwyn Brooks. 2011. *Here is New York*. 4th ed. New York: The Little Bookroom. Original edition, 1949.

YouTube. 2017. "Policies, Safety, and Reporting." *YouTube Help Website*. Accessed March 30, 2017. https://support.google.com/youtube/topic/2676378?hl=en.

Chapter 8

Revisiting privacy in public spaces in the context of digital vigilantism[1]

Daniel Trottier

Introduction

On the 22nd of June 2015, user JennyW22 posted two photographs on the popular image-hosting site, Imgur. These images depict a middle-aged woman walking down a sidewalk holding, among other things, two lawn furniture cushions. The post, entitled "Meet the Cushion Stealer!", indicates that the theft took place in downtown Kitchener, Ontario, Canada. The alleged thief is rendered visible through both rear and side perspectives, and is identifiable through her appearance and location. The images are accompanied by the following description.

> *This Women (sic) Stole From My Sister*
> This women (sic) stole the cushions from my sister's front porch in down town Kitchener, Ontario. A neighbour snapped some pictures as she walked away. If you see this women (sic), ask her to please return them, it's a stupid thing to steal and hurt my sister's feelings.[2]

The poster, identifiable through a user name that is not linked to any other posts on the Imgur platform, calls upon other users to identify and confront the woman. These other individuals may dwell in, or be familiar with, the city of Kitchener, although this call to action is reaching a much vaster network, which may in turn re-circulate it to contacts in Ontario, or Kitchener-based online groups. Although the poster is not explicitly calling others to provide any information about the identity of the target, such actions would be in line with her motivation to author this post. This type of collaboration is made possible through the comment feature, which is nearly ubiquitous in online social platforms. The posting garnered over 300 thousand views and 800 comments.

To be sure, this is a minor act of coordinated activity, and an otherwise mundane response to a petty theft. Yet it is also indicative of a rapidly normalized practice of responding to perceived criminal and moral offenses in public spaces, by using digital media technologies and the vast networks of individuals that they connect. User-led digital media enables the production and circulation of personal information, including photographs, film, and text-based descriptions of other people's behavior and appearance. Unanticipated and non-consensual use of this information has forced a reconsideration of privacy in public space. Continued development in the study of digitized privacy violations requires understanding the implications of how personal information about other individuals in public physical spaces following a perceived offense is shared through online platforms. While this kind of user-led surveillance has followed exceptional events such as riots (Schneider and Trottier 2012), it is becoming a prevailing method of engaging with legal and moral infractions in public space, and has become a troubling and often contested routine of contemporary life (Burns 2014). Here, the notion of public is expressed in a way that is both tethered to and decoupled from locale. Digital vigilantism involves circulation of information that renders a targeted individual visible on a vast geographic scale, such that the identity of a person engaged in mundane activity in a medium-sized Canadian town is accessible beyond that locale with greater scope and speed than could be provided by any other medium. Yet problematic local identities can be crafted, mobilized, altered, or misrepresented through these digitized platforms on a global scale. Moreover, the public designation of a scrutinized territory (such as a city street), coupled with the collective nature of a municipal, regional, or national identity can be mobilized in order to exclude as well as undermine the privacy and integrity of a targeted (category of) individual.

This chapter considers the extent to which recent assertions about privacy and digital media are also applicable to publicly accessible spaces in light of digital vigilantism. By revisiting Johnston's (1996) conceptualization of vigilante activity, the section below considers how mediated interventions can serve to augment the visibility of targeted individuals in public, and consequently compromise their privacy both within and beyond such spaces. Such considerations will contribute to a working typology of cases of mediated shaming of targeted individuals in public spaces. Digital vigilantism (DV) maintains conventional vigilantism's commitment to policing and securitizing a designated physical and public space, yet the nature of this citizen-led intervention is extended through mobile devices and social platforms. Consider, as an example, the emergence of so-called "crap parking" websites that serve to identify and shame offensive and potentially unsafe drivers (Moore 2018). While such groups appear to be primarily concerned with the municipality featured in their respective names, the capture, retention and circulation of identifiable details (such as license plates) through global mediated platforms suggest that the visibility enacted by being located in a public physical space at a specific point in time is augmented by these platforms. As the primary means of social harm employed is mediated, attempts to either maintain anonymity or violate privacy in public spaces are seemingly governed through digital platforms, and amplified by user activity. After reflecting on the changing nature of both vigilantism and privacy in light of digital media, this chapter identifies directions for subsequent research on mediated shaming of individuals in public spaces.

Vigilantism revisited

In order to speculate in which ways DV departs from conventional embodied vigilantism, we may return to Les Johnston's (1996) six elements of vigilantism and consider how each of these features is reinforced, augmented, or contested through digital media. First, there must be some degree of planning and premeditation on the part of instigators, such that an act of spontaneous self-defence would not be considered as vigilantism. While this remains the case with DV, digital media affordances greatly facilitate comparatively spontaneous coordination, such that material, spatial, and temporal barriers are obviated. While some research on digital media practices points to a degree of reflection and planning among users making social life visible (Davis 2010), the saturation of devices, platforms and their seamless integration with each other, combined with the embedded nature of a range of practices involving these tools, collectively function as a kind of quasi-automation of publishing of social and public life. With DV, there are pre-existing groups that fulfil the criteria of minimal planning, in that they publicly express a purpose and call upon members (and digital media users more broadly) to be on the lookout for offenses and offenders.[3] Yet, there are also spontaneous instances where, in the absence of a pre-determined planning, an individual may respond to an offense (e.g., racial or sexual harassment in the street) by photographing the targeted offender, uploading this photo along with a text-based description of their appearance and behavior. While the immediate responsive act is not formally planned, (a) it reflects a conscious and wilful act (and cannot readily be likened to Johnston's examples of a violent physical act committed in self-defense), and (b) that act triggers mediated actions among a broader public that may reflect some degree of planning and coordination.

Second, Johnston identifies private voluntary agency as a core requirement. In other words, citizens engaged in vigilantism must be distinguished from police and state actors, as well as from private entities that nevertheless "function within the legal ambit of the state" (Johnston 1996, 225). With DV, the relationship with police in many instances appears to be complicated, as the latter may appropriate data generated by DV efforts, as the Vancouver Police did following the 2011 riots (Leinwand Leger 2011). Additionally, police make appeals to the public for information that may trigger vigilante responses, as in the *Kopschopper* case in the Netherlands, where a group attack on a youth was captured by CCTV and circulated to the public. While the search for suspects was initiated by a state agency, what ensued was a user-led identification and mediated shaming of the suspects that diverged from what the police and courts found appropriate (Verschuren 2013). These developments are far from the first instance of police–public cooperation, yet digital media may facilitate a repurposing of police appeals or content for extrajudicial goals. In addition, private entities such as social platforms are arguably complicit actors in DV campaigns insofar as they facilitate such coordination. Vigilantism is defined primarily as activity conducted by private agents outside the remit of the state. Yet the fact that DV campaigns are more heavily mediated necessarily implicates a network of actors that include – among others – commercial social media platforms. Moreover, the cross-contextual nature of these platforms renders the boundary between police and public porous.

Third, conventional vigilantism is understood as a form of "autonomous citizenship" (Johnston 1996, 226) whereby citizens engage in self-protection within a given territory. As DV in practice connects embodied incidents (e.g., offending behavior or attributes within a given jurisdiction) with a communicative realm that exceeds that territory, we may consider that the notion of citizenship is complicated, though not necessarily ruptured.

Fourth, vigilantism involves the (threatened) use of force. While DV initiatives may contribute to embodied violence (for example, in the case of suspected child exploiters who are compelled to take their own lives (Krishnan 2016)), its key tactic is a form of cultural violence (Galtung 1990) that simultaneously renders targets visible, denigrates them on the basis of suspected actions as well as categorical membership (e.g., mistreatment of aboriginals in the context of municipal based groups in Canada), and legitimates accompanying forms of violence, including the structural foreclosure of life chances (Gandy 2009). The full potential for social harm inherent in the public circulation of an individual's image in an extrajudicial context is critically understudied in research on the practice of DV.

Fifth, Johnston notes that vigilantism "arises when some established order is perceived to be under threat from the transgression (or potential transgression) of institutionalized norms" (Johnston 1996, 229). Once again, with DV the ability to oversee other geographic and cultural contexts greatly complicates the practice of asserting a singular established order. In other words, it is possible that DV is not manifest as a matter of policing a local public space with local norms, but rather a fusion of localized publics with cross-contextual information sharing practices and norms that are manifest on digital platforms. In the case of groups based on an embodied physical space, such as "The Judge Beauce" (Myles and Trottier 2017), we may speculate about the extent to which its viewers and participators are concerned strictly with Beauce as a municipality, and in what ways their understanding of this threat is shaped by other similar groups for other regions (such as "Creep Catchers" in Canada (CBC 2018) or "Letzgo Hunting" in the United Kingdom (Barford 2013)), alongside a broader criminal media culture including the American television program *To Catch a Predator* (Kohm 2009). Qualitative interviews with members of such groups may illuminate how understandings of "the local" and "the public" are constituted.

Sixth, vigilantism serves to assert personal and collective security by providing "the assurance that an established system of order will prevail" (Johnston 1996, 231). This may remain the case with DV, yet we may question how the boundaries of the collective are asserted. In particular, the fact that information about offending acts is accessed and rendered meaningful through mediated channels, in turn rendering embodied spaces and territories visible and knowable, may be understood by participants to be a source of security. In other words, publicly accessible physical spaces may be understood to be secure through their visibility on digital media.

DV is represented in news media as a series of high-profile incidents such as riots (Mann 2011) or terrorist attacks (Madrigal 2013), yet it is also connected to a broader tendency that is informed by media logics and manifest as a set of practices reflecting one or many features of an ideal type. DV is a form of mediated and coordinated action. Although an empirically and conceptually rigorous typology of cases is needed, the majority of cases publicized in news media derive from perceived offenses that occurred in public spaces such as trains (Efstathiou 2016) and urban roadways (Chulkovskaya 2016). Some exceptional cases involve offenses that may occur behind closed doors, but nevertheless mobilize the public through the violation of commonly held fundamental values (as in the case of child exploitation, see for example Evans 2003; Williams and Thompson 2004). In response to this offense taking, users seek to render a targeted individual or category of individual visible through information sharing practices such as assembling and publishing their personal details, including their name, age, physical descriptions, but also photographs and videos of the offending acts, alongside other available information such as their employer ("doxing"). This response is typically led by individuals who may temporarily coalesce, for example, on a Facebook group, but are otherwise unaffiliated with a formal organization. DV campaigns are driven by a range of criminological motivations, including responding to criminal events as well as the prevention and deterrence of potential transgressions. For example, the 2011 Vancouver Riot Facebook group justified its coordinated action as identifying suspected rioters, but also claimed "this page is as strong a deterrent you will find to prevent this from happening again" (Vancouver 2011). Yet DV campaigns also pursue informational goals such as the identification of a targeted individual or category of individual, as well as articulating an understanding about shared norms and values. Consequently they can express a mediated collective identity that may be informed by national, religious or ethnic forms of solidarity, even if participants themselves are not physically proximate to one another.

As a starting point towards a more empirically and conceptually informed typology of digital vigilantism cases, it is already possible to consider a range of exemplars that are manifest in public spaces. The most prominent category is campaigns that emerge in response to exceptional events such as riots (Schneider and Trottier 2012) or acts of terrorism (Starbird et al. 2014). In these cases, users spontaneously create group pages on popular social media platforms in order to identify perpetrators. In addition to information gathering, these groups serve as opportunities to express grief over the offending acts, and exert a collective identity. These are events that occur in public spaces, and also constitute a public emergency, and as such the visibility conferred upon targeted suspects may be largely justified (although concerns over privacy, proportionality and the general appropriateness of such campaigns may still be expressed in group discussions).

A second category consists of groups that address comparatively minor social infractions occurring in public. These include so-called "crap parking" Facebook groups as well as Tumblr accounts dedicated to documenting and discussing the gendered use of space on public transit (Rose 2016). These campaigns involve similar platforms as the above examples, although these are established and managed in a premeditated manner (as they are set up in order to solicit information about forthcoming offenses and offenders). Conversations taking place in these groups not only cover the offending acts in question, but also the perceived appropriateness of documenting and circulating such offenses. In particular, the decision to identify the suspected target through highly publicized channels entail considerable privacy implications for those under scrutiny. Tentatively, it is possible to observe emerging protocols for documenting offenders, such as the decision to include the face of offenders, or to blur out license plates.

A third category of publicly oriented DV groups includes digital media groups that are based on a particular region or municipality. These groups often bear disparaging names such as "Thunder Bay Dirty",[4] and serve to document undesirable and dysfunctional aspects of the municipality, while also providing a venue to comment on these concerns. They are typically created and managed by a specific individual, yet encourage both feedback and original posts from group members, who may be presumed to have some connection to the city in question. While these groups do not express a specific DV-based agenda in their description, in practice they frequently include postings of individuals in public spaces who are designated as strangers and a public nuisance. Further research ought to consider this particular category of group as a mediated reproduction of municipal identities online. In particular, these groups can be considered an expression of civic and political grievances that are, in turn, aligned with emerging responses and retaliation that involve the strategic mediated visibility of targeted persons.

Finally, as seen in the introductory example, individual social media users may post details about public offenses they detect or witness on their own accounts instead of on designated groups. As an example, consider the case of a woman who encountered road rage and racial abuse at the hands of a Pepsi truck driver.[5] Again, such postings are usually spontaneous, as content about offending acts can be uploaded while the user is in the vicinity of the public offense. These are the most atomized category of DV, as in the first instance users are circulating information through their own ego-centric network rather than a broader public, depending on their profile's privacy settings. Yet, in the above case, information about the offense and offender were further distributed by a prominent political activist on Facebook, thus reaching a broader user base. The ease with which this information can circulate beyond those known to the poster, including re-posts, remediation, and leaks to other platforms, speaks to the potential of this content to "go viral" and reach a much wider public.

By virtue of the nature of the offending acts and the motivation of DV participants as well as other social actors such as the news media, the former would generate a greater amount of content and visibility in a comparatively limited period of time, but this content may quickly be removed from social platforms due to legal and ethical reasons (Reddit 2015). However, news media coverage will likely remain, including information about individual targets and their alleged offending acts.[6] In contrast, campaigns that target comparatively mundane offenses may exhibit all or many of the same characteristics of vigilantism, but with participants and critics expressing less concern over privacy of the target and proportionality of the intervention in response to the perceived offense. Although the distinctions between campaigns in response to major and minor offenses remains critically understudied, a pressing concern is that the routinization of low-level forms of DV may not readily raise concerns about privacy, and may therefore become a common means for citizens to police and render visible other citizens. This may be due to the perceived petty nature of the offending acts, such as minor theft or traffic infractions, which may influence perceptions of the severity of the punitive nature of the ensuing visibility. Another factor to consider is the often-marginalized status of many targeted individuals (such as aboriginals in municipal-based groups like "Thunder Bay Dirty" or women in public-transit groups like "Women Who Eat on Tubes"), which may preclude concerns about their integrity and human rights. Finally, the expression of many DV events as localized may lead those who engage in it as a more restricted form of local gossip, even if the total number of viewers of uploaded content may eclipse the entire population of a city like Kitchener, Ontario.

In considering the emergence of DV from conventional vigilantism, it appears that the further mediatization of such campaigns imposes a number of consequences, notably in terms of how citizenship and policing are re-articulated in contexts that merge localized environments with deterritorialized platforms (Harvey 1989). Moreover, the increasingly mediated nature of vigilante interventions, coupled with the ability to exceed any localized context, raise significant concerns about the potential visibility of the target, and the ensuing privacy challenges. What follows is an overview of privacy in the context of digital media, with an emphasis on key assumptions made (and frameworks mobilized) when discussing privacy in the context of digital media, and their applications to establishing a research agenda for DV.

Privacy and online shaming of people in public spaces

Privacy is central to both the study and personal experience of visibility through digital media. It is a legal concern linked to individual and collective rights, but is also performed through everyday social interactions. In practice, individuals have various conceptual models of privacy that are balanced against other priorities. Privacy's relevance in modern law was first identified as "the right to be let alone" (Warren and Brandeis 1890), that is, freedom from surveillance and scrutiny. A more contemporary legal interpretation of privacy rests on "the claim of individuals, groups, or institutions to determine for themselves when, how, and to what extent information about them is communicated to others" (Westin 1967, 337). This definition implies a degree of control over personal information flows that has been greatly complicated by digital media platforms, most notably citizen-led practices that render others visible through the gathering and circulation of their personal information.

A micro-sociological understanding of privacy is informed by Erving Goffman's work on self-presentation (Goffman 1959). His dramaturgical approach underscores the performative nature of social interactions by framing social life in terms of roles, stages, and audiences. Goffman argues that individuals are deliberate and strategic in their social interactions. This is partly for the sake of cooperating with others to maintain a cohesive understanding of the world, but also to remain in good standing with others. He distinguishes between front stages, locations, and contexts where social performances occur, and back stages where performances are managed. These regions map along a cursory understanding of "the public" and "the private," and are embedded in architectural designs (for example, most service sector workplaces have a clear barrier separating front and back stages), as well as digital media interfaces, where the distinction may be more fluid.

The construction and performance of identity gains salience when considering another one of Goffman's sociological contributions: stigma (Goffman 1963). Stigma refers to personal attributes that an individual wishes to hide from others, ranging from biographical details, physical traits, or social attributes. Goffman contends that individuals take great measures to hide stigmatizing information from public scrutiny, and that the privacy of the back stage is paramount for concealing these stigmas. Moreover, stigmas are as ubiquitous as the technologies and processes that risk uncovering them. Goffman raises the possibility that virtually any individual is associated with some personal information that is at best subject to shaming and discredit, and at worst liable to trigger (populist or state-led) legal intervention. In relation to DV, this suggests a potential ubiquity of personal information that could trigger a moral offense (in turn triggering a DV campaign), should this discrediting information be rendered visible. The growing digitization of personal records, coupled with the pervasive possibility of hacks and leaks, renders this an all-too-real possibility. Moreover, the search for additional personal information about targets in the midst of a DV campaign also raises the possibility of revealing such discrediting information, furthering the potential harm of those who have been publicly shamed. An oft-repeated critique of privacy concerns is that if you have nothing to hide, you have nothing to fear. Based on the pervasive nature of categorical discrimination, Goffman stresses that social interactions are typically a high-stakes affair.

Applied to DV, the notion of privacy (in terms of behavior that may take place in the private realm, and presumably shielded from public scrutiny) may support the view that offenses occurring in public (such as a city street) justify public intervention, as manifest through the production of photographs of the infraction, as well as circulation of these and supporting information alongside the circulation of additional personal details that have been later uncovered. Yet, even if we accept that it is reasonable to be photographed by others in public (although legal and cultural frameworks diverge on this point), the transition from momentary visibility in a public street, and being visible through publicly accessible and searchable platforms, including the first page on a cursory Google search of one's name, are by no means equivalent.

Users on contextually broad platforms like Facebook or Twitter encounter this challenge as they struggle to find information they are comfortable sharing with friends, co-workers, family, and others simultaneously. Multi-contextual services are compelled to develop privacy settings that are robust enough to maintain contextual boundaries. Perhaps the most important contextual violation has been the transition from online to offline. For instance, if the contents of a seemingly anonymous blog were recirculated in the author's work environment, this would be both the rupture of an expectation of privacy, and also the transgression from one contextual environment to another. In the context of DV, we are beginning to see the opposite violation: a perceived transgression (such as a petty theft in a public space) is recorded and uploaded to an online platform. Once digitized, it may circulate to a multitude of other contexts.

Popular discourse suggests a paradoxical distinction between a concern for privacy in principle and a willingness in practice to routinely disclose personal information (about one's self or another) online (Barnes 2006), especially when the practical implications are directed at a stranger who just committed a moral offense in a public space. Moreover, peers play a fundamental role in maintaining or endangering one's privacy, most notably during doxing campaigns. So-called "Friends" and "Followers" may have privileged access to a target's personal information, and thus bypass any technical assurance of privacy when re-circulating these details. In addition, some platforms, like Snapchat, were expressly designed for ephemeral communications (Poltash 2013), while others, like Twitter and Instagram, are primarily framed as tools for public expression. Yet, even Snapchat's claims of ephemerality have been compromised by users intentionally leaking private data to the public, alongside the discovery that the platform itself retained photos that were meant to be deleted (FTC 2014). Note that individuals are not typically bound to a single platform, and may choose one over the other at a particular moment on the basis of perceived privacy affordances. Moreover, in practicing DV, they may wilfully take steps that exceed reasonable expectations of privacy on a given platform, as part of a broader attempt to render a target socially visible.

User understandings of privacy are partially shaped through platform engagement, particularly default privacy settings (Boyd and Hargittai 2010). However, users adopt a variety of privacy-protective behaviors that are independent of privacy settings, such as relying on pseudonyms, as well as maintaining multiple accounts that may distinguish professional roles from personal ones (Trottier 2012). Several platforms and service providers such as Facebook and Google have explicitly targeted these practices through adopting "real name" policies. User tactics also include norms and expectations in embodied contexts; for example, the expectation that house-party attendees will not be photographed and tagged online. While such norms may come to take hold as digital media is further embedded in social life, public spaces will likely remain locations where it is justified for citizens (to say nothing of states and corporations) to photograph and identify individuals. Yet, in terms of risks calculated when negotiating an online presence, few users currently anticipate the targeted scrutiny of DV or doxing campaigns. Indeed, popular accounts of such campaigns indicate that these are largely unanticipated (Ronson 2015). We may speculate that the pervasiveness of such campaigns may shape public concerns, or at least bring about a re-imagining of the potential visibility of one's presence in a public space, and the myriad of platforms and users who may access, comment on, contribute to, and act upon this information.

Privacy has become a salient term to discuss online shaming of conduct in public spaces. Yet, thinking about high-visibility shaming exclusively in terms of privacy limits the scope of a social-scientific analysis and raises a number of concerns. First, privacy is typically conceived as an individual matter, typically overlooking precisely how the social dimensions of social media may put other social actors in harm's way. User privacy, often framed as an individual responsibility, and privacy settings for social media as well as public recommendations by advocacy groups support the assumption that it is the user's responsibility to govern their own data. Yet, control over what personal information is knowable to what social actors is a form of autonomy that is largely beyond individual control. As seen in the case of JennyW22, a suspected cushion thief is made visible through the acts of another individual who was physically present to photograph the woman in public.[7] This content was further circulated and commented on by other digital media users, and may in turn shape the suspected cushion thief's public engagements in the streets of Kitchener. Through posts, tags, and comments, digital media users routinely disclose information about their peers, exposing others to public scrutiny and shaming. Likewise, the decision to photograph strangers in public spaces and upload this content is fundamentally a matter of exerting control over their visibility and reputation, especially when these are accompanied by calls to identify and shame that targeted individual. Concern for the self should extend to a concern for how others are subject to exposure through our own actions.

Related to this individualistic bias in privacy discourse, managing one's own privacy often results in the increased scrutiny of others. Individuals will understandably take measures to safeguard their privacy, notably in the maintenance of private spaces that necessitate securing boundaries, carefully monitoring those who enter and exit that space. Consider the gated community equipped with security cameras and facial recognition software, or the private social media platform that requires visitors to identify themselves and make their actions visible while interacting with others. Likewise, upon finding out that their privacy could be compromised on social media, young users interviewed in an earlier study reported that they began watching over content their friends would post, taking active steps to manage or sanction these friends (Trottier 2012). These measures exemplify the internalization of the surveillant gaze (Foucault 1977) applied outwards to peers, in the name of self-preservation. We may consider whether the calculation of generalized risk and security in public spaces (that is not entirely distinct from preservation of one's own privacy and integrity) compels a heightened vigilance whereby citizens are willing to inflict violence against others by rendering them visible in a criminalized framework.

A further limitation of privacy is that it is too easily conceived of as a resource or commodity, which means that those with greater capital or purchasing power will be able to make use of privacy, while others will simply have to cope with unwanted and unanticipated exposure. Privacy thus becomes a luxury good instead of a fundamental right (Papacharissi 2010). In defining privacy as the absence of embarrassing and harmful forms of exposure, it is evident that detached houses afford residents more privacy than apartments, and first-class airport lounges are a comparatively safer space than a crowded concourse in which to commit a jetlag-fuelled gaffe. In the context of public spaces, we may suppose that those who lack privileged access to private spaces are especially vulnerable to DV, not only through their heightened presence in public/visible spaces, but also through their membership of categories of suspicion such as "at-risk" or "anti-social." These categories may in turn preclude their participation in quasi-public spaces like shopping malls and airports. In terms of digital media, navigating interfaces and combing through content requires particular sets of media literacy, and also costs time, to say nothing of hardware and services. Many users simply do not have enough of either, and will be disproportionately subject to invasive scrutiny. Public relations efforts on social media are informed by high-profile cases involving politicians, celebrities and other public places. This arguably constitutes a normalization of strategic identity management, along with a proliferation of online reputation management services and consultancies. While legal debates and political advocacy focus on establishing fundamental privacy and data protection rights and legislation, commoditizing privacy can circumvent a collective and rights-based preservation of privacy.

Discussion

Through digital tools, vigilantism is becoming manifest as mediated visibility and ensuing social harms. These practices extend the scope of visibility of targeted individuals in public spaces, along with other facets of their private lives that may be otherwise partly visible online and accessed during doxing campaigns. DV campaigns concern both exceptional and mundane activities and as such are dispersed in their motivations and degree of severity. Along with sociotechnical practices that are harnessed as a means to harm the social standing of others, norms of acceptable conduct are formative at this stage. In engaging in (or responding to) DV, users are mobilizing several discourses about, among other things, privacy and visibility. Concerns for proportionality and rights of the individual are often invoked, yet these may be set aside in response to exceptional crimes, or simply overlooked in the case of mundane breaches of social norms. The distributed and viral nature of mediated visibility complicates any attempts to establish an agreeable and achievable understanding of proportionality (DV as means to apprehend and deter an offender): not only does personal information about a target circulate among an often immeasurable and uncontrollably scaled audience, but the range of personal information about the target (and their family, workplace, etc.) is also inestimable.

In outlining this growing field of activity, we can identify three points of concern over privacy and the governance of publicly accessible spaces, which may provide indications for subsequent data collection and analysis. First, the practice of DV is a means of leveraging several dimensions of "the public," including (a) an alleged public opinion or judgement about an offending act (regardless of how marginal that opinion may be), which (b) itself has occurred in a public space (which for the purposes of this chapter refers to embodied spaces, but may extend to mediated publics), and (c) in which the denunciation and information circulation itself is occurring through publicly accessible channels. Subsequent research should consider how each of these dimensions of "the public" stand in relation to privacy, and in particular if articulations of public interest through each of these dimensions is leveraged to supersede the rights of the private citizen under scrutiny. In particular, the articulation of a collective identity and perceived moral legitimacy within a given spatial territory would appear to serve as a catalyst towards privacy violations.

Second, in considering the long-term ramifications of DV campaigns, it would appear likely that social harm to the targeted individual exceeds the period of public denunciation and information circulation, which complicates any attempt at proportionality in instances of online public shaming. Once personal details about the target have spread, they may remain accessible through search engines whose ranking criteria privilege content that has experienced a viral "burst" at one point in time, which may in turn have some bearing on the target's reputation and public standing. Yet, while targets are today searchable through their own name, or the designated name of the (category of) incident in which they are implicated, there is the growing probability of being identified through facial recognition, or other means of building links between hitherto unstructured data. Such possibilities can rest on both the automated processing of data and crowdsourcing efforts that mobilize individuals who vary in their technical skills and socio-political motivations. This would suggest that the mediated and augmented visibility of the public street might emerge in job interviews or border crossings, among other moments. Such possibilities implicate other concerns that are related to and distinguished from privacy violations, including a chilling effect on online speech among potential targets, but perhaps also a chilling effect on presence and visibility in public. Subsequent research should consider how targets of DV negotiate public spaces (embodied and mediated) during the supposed afterlife of their campaigns. Here we may also consider those who are rendered visible through their own participation in the targeted shaming of others, and how they may negotiate this visibility through, for example, the decision to participate anonymously or with accounts linked to real names.

Finally, while vigilantism is commonly framed in terms of extrajudicial interventions, the conceptual boundary distinguishing state-sponsored and citizen-led forms of policing warrants empirical investigation, especially as high-profile cases of DV have been triggered by police intervention (*RTL Nieuws* 2013). Police agencies may wish to utilize forms of crowdsourcing and nodal governance, and will likely negotiate effective forms of data handling as a means to prevent or dampen unsanctioned citizen-led interventions. Focusing on these negotiations will undoubtedly generate insight about how state-led and private security agencies understand the aforementioned dimensions of the public in the context of distributed forms of information gathering.

Notes

1 This work was supported by the Netherlands Organisation for Scientific Research (NWO), project number 276-45-004.
2 http://imgur.com/gallery/P4Vhf/.
3 See for example the Facebook Group "Women Who Eat on Tubes" (https://www.facebook.com/groups/WWEOT/) as well as the response group "Men Who Photograph Women Who Eat on Tubes" (https://www.facebook.com/groups/654841407920650/).
4 https://www.facebook.com/Thunder.Bay.Dirty.2.0.
5 https://www.facebook.com/shaunking/photos/a.799605230078397.1073741828.799539910084929/1135217836517133/.
6 See for example press coverage of accused participants in the 2011 Vancouver Riot (Ryan 2011).
7 It is worth noting that as the photographs were reportedly taken from inside the neighbors' house, neither the photographer nor JennyW22 were 'in public' in the above sense.

References

Barford, Vanessa. 2013. Who are vigilante group Letzgo Hunting? *BBC*, 19 September. www.bbc.com/news/magazine-24143991.

Barnes, Susan B. 2006. A privacy paradox: Social networking in the United States. *First Monday* 11(9). http://firstmonday.org/article/view/1394/1312.

Boyd, Danah and Eszter Hargittai. 2010. Facebook privacy settings: Who cares? *First Monday* 15(8). http://journals.uic.edu/ojs/index.php/fm/article/view/3086/2589.

Burns, Matt. 2014. Google explains how not to be a glasshole. *TechCrunch*, 18 February. https://techcrunch.com/2014/02/18/google-explains-how-not-to-be-a-glasshole/.

CBC. 2018. Alberta Creep Catcher gets fine, probation for "repugnant and malicious" crime. *CBC News*, 22 January. www.cbc.ca/news/canada/calgary/red-deer-creep-catcher-carl-young-sentence-fine-probation-1.4498940.

Chulkovskaya, Yekaterina. 2016. StopHam: The end of the road for Russia's parking vigilantes? *Russia Beyond The Headlines*, 4 April. http://rbth.com/politics_and_society/2016/04/04/stopham-the-end-of-the-road-for-russias-parking-vigilantes_581711.

Davis, Jenny. 2010. Architecture of the personal interactive homepage: constructing the self through Myspace. *New Media & Society* 12(7): 1103–1119.

Efstathiou, Zoe. 2016. Shamed businessmen DENY claims they "refused to give up seats for disabled woman". *Daily Star*, 23 September. www.dailystar.co.uk/news/latest-news/547395/Disabled-woman-furious-after-businessmen-refused-to-give-up-her-reserved-seats-on-train.

Evans, Jessica. 2003. Vigilance and vigilantes: Thinking psychoanalytically about anti-paedophile action. *Theoretical Criminology* 7(2): 163–189.

Foucault, Michel. 1977. *Discipline and Punish: The Birth of the Prison*. London: A. Lane.

FTC. 2014. Snapchat settles FTC charges that promises of disappearing messages were false. *Federal Trade Commission Press Releases*, 8 May. www.ftc.gov/news-events/press-releases/2014/05/snapchat-settles-ftc-charges-promises-disappearing-messages-were.

Galtung, Johan. 1990. Cultural violence. *Journal of Peace Research* 27(3): 291–305.

Gandy, Oscar H. 2009. *Coming to Terms with Chance: Engaging Rational Discrimination and Cumulative Disadvantage*. Farnham: Ashgate.

Goffman, Erving. 1959. *The Presentation of Self in Everyday Life*. New York: Anchor Books.

Goffman, Erving. 1963. *Stigma: Notes on the Management of Spoiled Identity*. New York: Simon & Schuster Inc.

Harvey, David. 1989. *The Condition of Postmodernity*. Oxford: Blackwell.

Johnston, Les. 1996. What is vigilantism? *British Journal of Criminology* 36(2): 220–236.

Kohm, Steven A. 2009. Naming, shaming and criminal justice: Mass-mediated humiliation as entertainment and punishment. *Crime, Media, Culture* 5(2): 188–205.

Krishnan, Manisha. 2016. An Edmonton woman killed herself after being confronted by creep catchers. *Vice.com*, 29 September. www.vice.com/en_ca/article/an-edmonton-woman-killed-herself-after-being-busted-by-creep-catchers.

Leinwand Leger, Donna. 2011. Social media aid Vancouver police in identifying rioters. *USA Today*, 23 June. http://usatoday30.usatoday.com/tech/news/2011-06-22-facebook-vancouver-riots_n.htm.

Madrigal, Alexis C. 2013. Hey Reddit, enough Boston Bombing vigilantism. *The Atlantic*, 17 April. www.theatlantic.com/technology/archive/2013/04/hey-reddit-enough-boston-bombing-vigilantism/275062/.

Mann, Bill. 2011. Social media "vigilantes" I.D. Vancouver rioters – and then some. *The Huffington Post Canada*, 2 July. www.huffingtonpost.ca/bill-mann/vancouver-riot-social-media_b_889017.html.

Moore, Charlie. 2018. Are these Australia's worst parkers? Facebook page dedicated to shaming "road muppets" returns – and the pictures will INFURIATE you. *DailyMail.co.uk*, 28 February. www.dailymail.co.uk/news/article-5445249/Are-Australias-worst-parkers.html.

Myles, David and Daniel Trottier. 2017. Leveraging visibility, gaining capital? Social media use in the fight against child abusers: The case of *The Judge Beauce*. *Social Media + Society*. (online first) https://doi.org/10.1177/2056305117691998.

Papacharissi, Zizi. 2010. Privacy as a luxury commodity. *First Monday*, 15(8). http://journals.uic.edu/ojs/index.php/fm/article/view/3075/2581.

Poltash, Nicole A. 2013. Snapchat and sexting: a snapshot of baring your bare essentials. *Richmond Journal of Law & Technology* 29(4): 1–24.

Reddit. 2015. Promote ideas, protect people, 14 May. www.redditblog.com/2015/05/promote-ideas-protect-people.html.

Ronson, Jon. 2015. How one stupid tweet blew up Justine Sacco's life. *New York Times Magazine*, 12 February. www.nytimes.com/2015/02/15/magazine/how-one-stupid-tweet-ruined-justine-saccos-life.html?_r=0.

Rose, Rebecca. 2016. Meet the Orthodox woman behind 'Move the fuck over, Bro' blog. *Jezebel*, 15 January. http://jezebel.com/meet-the-orthodox-woman-behind-move-the-fuck-over-bro-1501688008.

RTL Nieuws. 2013. Welke straf krijgen de kopschoppers van Eindhoven? *RTL Nieuws*, 11 December.http://www.rtlnieuws.nl/nieuws/binnenland/welke-straf-krijgen-de-kopschoppers-van-eindhoven.

Ryan, Denise. 2011. Rioting teen Nathan Kotylak and family face backlash, forced to leave home. *Vancouver Sun*, 11 June. www.vancouversun.com/technology/Rioting+teen+Nathan+Kotylak+family+face+backlash+forced+leave+home/4972283/story.html.

Schneider, Christopher and Daniel Trottier. 2012. The 2011 Vancouver Riot and the role of Facebook in crowd-sourced policing. *BC Studies: The British Columbian Quarterly* 175 (Autumn): 57–72.

Starbird, Kate, Jim Maddock, Mania Orand, Peg Achterman and Robert M. Mason. 2014. Rumors, false flags, and digital vigilantes: misinformation on Twitter after the 2013 Boston Marathon Bombing. *iConference 2014 Proceedings*, 654–662. doi:10.9776/14308.

Trottier, Daniel. 2012. *Social Media as Surveillance: Rethinking Visibility in a Converging World*. Farnham: Ashgate.

Vancouver. 2011. *Vancouver Riot Pics: Post Your Photos*. www.facebook.com/vancouverriotphotos.

Verschuren, Corné. 2013. Rechter laat tonen bewakingsbeelden zwaar wegen in Eindhovense 'kopschopperszaak'. *Omroep Brabant*, 28 August. www.omroepbrabant.nl/?news/1984791483/Rechter+laat+tonen+bewakingsbeelden+zwaar+wegen+in+Eindhovense+%E2%80%98kopschopperszaak%E2%80%99.aspx.

Warren, Samuel and Louis D. Brandeis. 1890. The right to privacy. *Harvard Law Review* 15(5): 193–220.

Westin, Alan. 1967. *Privacy and Freedom*. Atheneum.
Williams, Andy and Bill Thompson. 2004. Vigilance or vigilantes: The Paulsgrove Riots and policing paedophiles in the community Part II: The lessons of Paulsgrove. *The Police Journal* 77(3): 193–205.

Chapter 9

Emergency calls with a photo attached

The effects of urging citizens to use their smartphones for surveillance

Gerard Jan Ritsema van Eck

Introduction

With the rising popularity of smartphones, citizens are being offered increasingly fast mobile internet connections with exceedingly high or even non-existent data caps. At the same time, the quality of smartphone cameras is increasing rapidly. This means that, at least from a technical point of view, any citizen on the street could record high quality photographs or videos of emergencies and share them with the police instantaneously.

However, most emergency call centers are geared towards receiving phone calls, not photos and videos, from the public.[1] This means that, in situations where time is of the essence, all the capabilities that modern smartphones offer are suddenly rendered useless. To remedy this situation, various market parties, predominantly from the United States of America (e.g. RapidSOS 2016), have recently started offering configurations that make it possible to receive and process reports based on a wider variety of media. Besides market parties developing standalone "instant police reporting applications" and the related infrastructures, federal institutions on both sides of the Atlantic are supporting similar projects. Most notable is a push to make the 911 emergency response system in the United States capable of allowing "digital information (e.g., voice, photos, videos, text messages) to flow seamlessly from the public, through the 911 network and eventually, directly to first responders" (The National 911 Program 2016); and an effort within the EU Horizon2020 research program to create "a solution including a new smartphone app and on-line portal which are capable of being deployed in any European city" (Community Research and Development Information Service 2015a).[2] Finally, many police forces are exploring how they might leverage already established communication platforms, such as WhatsApp or Facebook, although doubts remain about the feasibility of using them for emergency situations (Nederlandse Omroep Stichting 2016).

Object of the current chapter

These developments should raise scholarly concern from those working in the field of surveillance studies. Each new configuration of surveillance capabilities introduces a new configuration of actors and their capabilities; opening up new strategies, new vulnerabilities and new avenues for resistance (Lauer 2012; de Certeau 1984). Interestingly, the smartphone enters the academic debate on surveillance mostly as an object which can be surveilled by (state) authorities or as a tool for sousveillance (e.g., Reilly 2015; Timan and Albrechtslund 2015, 4; Leistert 2013). This is in line with the – until recently – minimal state efforts to engage citizens to actively use their smartphones to collect data on ongoing incidents. Additionally, those who did record such footage "did not feel a part of a surveillance network" (Timan and Albrechtslund 2015, 16). This silence comes despite a wide range of possibilities that the ubiquity of smartphones presents to policy makers looking for new surveillance tools.[3]

The object of this chapter, therefore, is to investigate the societal effects of instant police reporting applications for smartphones. Not only does this fill a lacuna in the current literature on surveillance and smartphones, it shows that such possibilities come at a cost: processing and storing the gathered information leads to a loss of privacy in public spaces, and exhorting citizens to act as police informants introduces discriminatory biases, pressures towards normalization, and chilling effects.

Overview of the current chapter

Theories from surveillance studies will be used to explore the societal effects of instant police reporting applications. The theoretical developments in the field of surveillance studies can be divided into three phases: panoptical, post-panoptical, and contemporary conceptualizations which focus on thorough applications and refinements of the theories developed in the previous two phases (Galič, Timan, and Koops 2016). This chapter falls squarely in the latter category.

In the first section of the theoretical body of this chapter, entitled "Societies of discipline," panoptical theories will be used to analyze the effects of instant police reporting applications. In the section entitled "Societies of control," post-panoptical theories will be applied. Since the focus of this chapter is on societal effects, a third section is dedicated to discussing the implications for policy makers before turning to the conclusion.

A note on terminology

Before engaging with the theoretical body of this chapter, it is necessary to delineate the exact phenomenon under scrutiny. Instant police reporting applications have not taken on any definitive form yet and, as such, no more than a working definition can be provided. For the purposes of this chapter, a preliminary definition of the term "instant police reporting app" (or application) will therefore be used to denote any scenario in which:

i a person acting in a private capacity uses a mobile computing or communication device (which is usually a smartphone); to
ii communicate to a police control room data other than (only) voice or text data; about
iii a current or very recent situation; for which
iv quick police intervention is deemed necessary by said person, either to deescalate/end the situation or to apprehend any alleged transgressors (whether or not they have fled).

The data referred to in the second point could, for instance, be images (either still or moving[4]), audio files, geolocation data, or combinations thereof. There may be no dedicated stand-alone mobile application involved in any step of such a scenario, as such communications might also take place in other software environments – for example, e-mail or instant messaging apps. The term instant police reporting apps is used to cover all such scenarios to ensure consistency.

Societies of discipline

Michel Foucault (1995) was the first to describe the panopticon as a schema for the functioning of power. He based much of his analysis on Jeremy Bentham's infamous prison plans. In the short term, panoptic architectures endeavor to make their objects (be they prisoners, pupils, or the population as whole) potentially visible and thereby vulnerable to disciplinary intervention when they engage in behaviors deemed undesirable by the operators of the panopticon. In the long term, this vulnerability is internalized by the objects of the panopticon, consequently turning them into normalized "docile bodies." Foucault argued that power could function according to this efficient diagram not only in enclosed settings, but even in society at large. Such a generalized mode of social control is typical of what Foucault termed societies of discipline (Foucault 1995, 209).

Foucault's ideas of the panopticon and panopticism became standard references in the field of surveillance studies. In recent scholarship, panoptical theories have been especially prevalent in studies on closed-circuit television (CCTV) systems[5] because they share a central working mechanism with the panopticon: a visibility to the gaze of invisible watchers. This "surprisingly similar" manner of operation suggests that the panopticon provides a valuable theoretical model for understanding visual surveillance in the contemporary city. However, various critics hold that there might be more suitable metaphors (Koskela 2003; especially page 293, where she gives a more detailed treatment of some of the critiques).

The central question in this section is, therefore, whether and how the panoptical scheme can provide a deeper understanding of the working mechanisms of instant police reporting applications.

This section is divided into three subsections. The first subsection begins by investigating how smartphone applications for instant police reporting relate to panoptical theories in the short term and compares this to descriptions of the working mechanisms of CCTV in the academic literature. The second subsection studies the long term effects. In the final subsection, attention will turn to a possible empowerment of the users of instant police reporting applications in the panoptical diagram.

Short term effects: visibility

In a typical-use scenario of instant police reporting applications, the user would snap a photograph of a criminal act in progress, which would then be transmitted to a police control room and then retransmitted to officers on the street who might be able to immediately apprehend a suspect. This prompt police response is enabled by the visual nature of the material being shared: whereas a phone call to the emergency services would have resulted in the same flow of information, the nature of the information shared would be very different: rather than an image, a verbal or textual description of a suspect would be circulated.

The effectiveness of image sharing can be exemplified by considering the central figure in Figure 9.1, a photograph taken with a smartphone in the center of London. A fairly accurate description of the central figure would be that of a male of average height with short hair, wearing a black suit and backpack, listening to white earphones. Phoning in this description to the police would probably yield very little actionable intelligence given the probable prevalence of persons fitting that description in central London. However, if the photo in question were to be circulated to police officers patrolling in the vicinity of where the photo was taken,[6] the police might stand a much better chance at apprehending the suspect: his visibility renders him vulnerable to disciplinary intervention. What is thus most striking about the short term effects of instant police reporting applications in relation to the workings of a panopticon is this very direct link between the visibility of the suspect and her or his vulnerability to disciplinary intervention.

This mechanism is nearly identical to that of CCTV systems – perennial symbols for surveillance and the panopticon in particular. The efficacy of CCTV systems in crime reduction seems to be limited at best, however, for a number of reasons (see Welsh and Farrington 2002). Although CCTV *cameras* watch over many public spaces, the triviality of the bulk of the collected footage means that the majority of the video streams are only ever watched by *people* retroactively, and only if disturbances or crimes have been reported. In those cases where there are operators actively watching live camera footage at all times in the hopes of spotting a crime,[7] the monotony of the images leads them to engage in unproductive behaviors (le Goff, Malochet, and Jagu 2011, 45). So-called "smart" CCTV systems are one strategy to alleviate these shortcomings, incorporating algorithms that automatically alert operators to disruptions of a normal state of affairs.

Although instant police reporting applications also present a filtered data stream, there are two salient differences. First, the filter is formed by watchful citizens rather than by algorithms; this point will be returned to at length on p. 163. Second, the geographic reach of smartphones is limited only by that of their users, not by police budget constraints or bureaucratic and civic procedures regarding the placement of cameras. This means that the visibility of citizens vis-à-vis the police is heightened and geographically widened, and thereby the vulnerability to disciplinary intervention is increased.

Long term effects: normalization

The long-term effect of panopticism in Foucault's disciplinary societies is normalization. Over time, the knowledge that deviation is followed by punishment is internalized by the subjects and they no longer even consider deviating; the functioning of power has become independent of whether it is being exercised (Foucault 1995, 201). In this subsection, the focus will be whether the functioning of power through instant police reporting applications can become automatic enough to stop a potential transgressor from engaging in deviant behavior.

A crucial first step is convincing subjects that they are always potentially visible to authorities with the ability to punish. It is trivial to convince anyone in the industrialized world that wherever others are present, smartphones are also present. However, convincing anyone that a smartphone constitutes, in terms of its abilities to summon the police, not only a potential call to an emergency number but also an image recording and streaming device might be less straightforward.

Whether such claims can be made convincingly will depend on whether widespread uptake of instant police reporting apps will take place. If regulatory pressure develops to require phone manufacturers to make such functionalities available (as is currently the case with regard to voice-based emergency calls)[8], uptake would be nearly ubiquitous. However, regardless of such regulatory pressures, a part of the population might choose to frequently use instant police reporting apps, either because their occupations mean that they come into contact with illegal behavior frequently, or because they have a strongly developed sense of civic duty[9] combined with a strict set of (moral, social, and/or legal) norms which are thus regularly violated.

This suggests that the active userbase will not reflect society at large, but will be made up of a subset of the population who can not only afford a smartphone with a mobile broadband subscription, but also choose to actively

Figure 9.1 Photograph taken with an Apple iPhone 6 smartphone in London on 17 September 2015 by Benjamin Welle
Available at https://flic.kr/p/A81u4w under the Creative Commons Attribution-Non-Commercial 2.0 Generic license available at http://creativecommons.org/licenses/by-nc/2.0/.

participate. This means that the potential long term intensification of panoptical principles outlined above is only possible on the terms of these citizens.

An empowerment within the panopticon?

Rather than just an increase in the powers of the police (or, more broadly, the state security and safety apparatus), the reporting citizens are also endowed with certain powers by instant police reporting applications. This subsection will investigate how such an empowerment might take place and what the effects thereof might be.

Encouraging citizens to take an active role in panoptical schema is not a new idea. Bentham already opened the inner workings of his model prison to a scrutinizing public (Foucault 1995, 207). Groombridge has sketched the notion of an "omnicon", where the community, rather than a state agent, monitors the feeds from local CCTV cameras (Groombridge 2002, 43). This would shield against the excesses of surveillance because it is easier to differentiate between normal and abnormal behavior on the basis of the contextual knowledge that neighbors have about each other (Groombridge 2002, 42; Trottier 2014, 621). Groombridge's preliminary sketch has been expanded by Goold (2006, 12–14) with regard to how the footage from CCTV cameras in public places might be widely distributed and watched.

Because these analyses predate the advent of affordable smartphone cameras and mobile broadband internet, they assume a privileged position for the state concerning the placement of cameras. Thus, their focus is on how the *watching* of visual surveillance material might change, rather than on the mechanisms used for *collecting* such material. However, it is at the stage of collection where instant police reporting applications introduce new processes that can empower citizens.

Such an empowerment within the panoptic diagram should be problematized. Compare, for instance, two studies of CCTV operators who received no training in identifying suspicious behavior:[10] French CCTV operators described by le Goff, Malochet and Jagu (2011, 26–29) routinely engaged in discriminatory practices by concentrating disproportionally on (black) youth whose clothing and appearance signaled that they belonged to a "subversive" youth culture; British CCTV operators have shown similar behavior – being male, young, black, or a combination thereof made visitors to city centers more than twice as likely to be the target of the scrutiny of operators than could be expected on the basis of their prevalence in either the general population or arrests statistics (Norris and Armstrong 1999, 162–63).

Daniel Trottier (2014, 622) points out that there is little reason to believe that surveillance amateurs will engage in less discriminatory behaviors, especially if they are instructed to be vigilant for suspicious activity. Consider, for example, the case of nextdoor.com, a social networking site targeted towards supporting neighborhood communication. Despite its seemingly benign goal, the company has been battling accusations of facilitating racism, as many of the posts on the platform relating to security incidents seem to involve at least a degree of racial profiling (Harshaw 2015; Levin 2016). It seems that in a politico-societal context of responsibilizing "see something, say something" campaigns and moral governance towards personal risk reduction (Hier, Walby, and Greenberg 2006, 236), ordinary citizens will rely on stereotypes when given a broad mandate to select any scene they come across for further police scrutiny.

In conclusion, the functioning of instant police reporting can be panoptical. However, it is a panopticon that is partly run by, and on the potentially discriminatory terms of, the most frequent users.

Societies of control

The role of databases containing troves of potentially incriminating personal data is crucial for understanding any contemporary surveillance technology, a point to which Gilles Deleuze's (1992) "Postscript on the Societies of Control" draws our attention. It is related to two other topics which he points out and which are of concern for this section: database entries containing information on identified or potentially identifiable individuals; and the effects that storing large numbers of (bits and pieces of) such data in rhizomatically connected, temporally open-ended, and ever-modulating databases can have on society. The central question in this section is therefore what the societal effects of storing data obtained through instant police reporting applications might comprise, and who will be targeted by these effects.

To answer this question, theories that can broadly be labelled as "post-panoptical" (Galič, Timan, and Koops 2016) will be leveraged. In the first subsection, attention will be paid to whose data could be more likely to be included in databases as a result of instant police reporting. This question of addition to surveillance databases is fundamental; it determines who can be targeted and, conversely, who *cannot* be targeted by the possible societal effects to which attention will turn in the other subsections on anonymity, normalization, the presumption of innocence, and cumulative disadvantages resulting from simulation and prediction, respectively.

As will become clear throughout this section, the salience of post-panoptical theories depends on specific implementations and everyday uses that have not yet materialized at the time of writing. Therefore, this section needs to be more tentative than the previous section. Analogies with current surveillance practices are used where possible to infer possible future trajectories of instant police reporting applications.

Inclusion in databases

Surveillant assemblages, as introduced by Haggerty and Ericson (2000) and deeply influenced by Deleuze, are potentially endless multiplicities of objects and processes that form a functional unit for the purposes of surveillance. Such assemblages can be rhizomatically reassembled into ever-larger and modulating assemblages (Haggerty and Ericson 2000). However, a functional unit is needed which allows surveillant assemblages to be tied together: a database where the data about individuals who are deemed (at least) "of interest" are stored. In the next subsections, we will discuss the centrality of such databases in determining the societal effects of instant police reporting applications.

Instant police reporting applications generate images for immediate police usage. If these images are of a certain minimum quality, they also record those "gaseous flows of personal information" that bodies willingly and unwillingly emit (Smith 2016). An image of a body is thus also an image of the biometric characteristics of that body. Images therefor could facilitate the creation of (bits and pieces of) data about persons that is stored and analyzed in digital databases. By making use of metadata appended to either the image or the report, location and other information could also be extracted and stored.

Ostensibly, this could happen to anyone who happens to walk by in the background of an image snapped for instant police reporting purposes: a quick glance at Figure 9.1 reveals many innocuous bystanders.[11] However, a closer investigation of two of the dynamics at play reveals that specific groups of people have a greater chance of being included in surveillance databases as a result of instant police reporting applications.

First, the influence of biases held by the reporting citizens, which was already discussed in the subsection on empowerment within the panopticon, has recurring salience here. Being young, dark-skinned, male, or visibly belonging to some marginalized group might increase the risk of being included in databases as a result of instant police reporting applications.

Second is the influence of the selection of, and by, frequent users. When someone is spotted whilst stealing from a cash register, it might be obvious that s/he needs to be reported to the police as this clearly concerns a criminal act. But what about a drunken teenager? Should s/he be reported to the police? At what point does "drunk and disorderly" become a matter that should be brought to the attention of the police?[12] As was pointed out in the subsection on the long-term effects of the panopticon, the most avid users of instant police reporting applications will be those with a strong sense of civic duty combined with strict norms. Therefore, the thresholds for reporting and subsequent inclusion in databases in such ambiguous cases will be formed by the (moral, social, or legal) norms of those persons who have the *lowest* thresholds.

This means that those who frequently exhibit deviant behavior in public places will more frequently end up in police databases, whether the behavior in question was illegal or not. This will include those who unwittingly exhibit such transgressive behaviors: people diagnosed with mental disorders (including but not limited to, for instance, autistic or psychotic disorders), and "newcomers" to specific public spaces such as immigrants, young children, and tourists who are not yet aware of the local *mores*. Because these groups are already under heightened scrutiny from various (state) authorities, the added scrutiny as a result of instant police reporting apps can exacerbate earlier harms.

There are also those who wittingly exhibit transgressive behaviors: for instance, protestors – who will be discussed in the subsection on normalization and chilling effects.

Loss of anonymity vis-à-vis the police

An effect of the inclusion of data from instant police reporting applications in surveillance databases is that it might become easier for the police to identify who was doing what at a certain location, at a certain time, and with whom. This effect is presupposed in all of the other effects listed in the subsections below and it is thus crucial to better understand it before moving on. In general, identification of suspects is one of the reasons for surveillance in public spaces, and the reports from instant police reporting applications form nothing more than a new source of material in this regard.

In order to facilitate identification after an instant police report has come in, readily available software can translate image data to (biometric) data, store it, and compare it to previously collected data. In this way, an operator who receives an image of a suspect can have a quick answer to the question "Has this person been photographed previously?" If no positive identification can be inferred, perhaps a link to data collected earlier can be made, slowly but surely building a profile of an as-of-yet unnamed citizen.

Here the virtually endless potential for rhizomatic coupling of surveillant assemblages through the use of databases becomes apparent: if an algorithm can compile a database of (pieces of data about) individuals who may be of interest using images submitted through instant police reporting applications, a – separate but connected, or the same – database can also be populated using data from for instance smart CCTV cameras. Compare here the already long-established practice of holding on to "dumb" CCTV material for a limited period of time in order to be able search through them manually when a report comes in (le Goff, Malochet, and Jagu 2011, 41): this would merely lend that process a mathematical hand. Furthermore, local, national (e.g., the Next Generation Identification database of the FBI), and international databases of wanted persons would also form a logical addition here, as might databases maintained by private parties (Winston 2015).

To conclude this subsection, this new source of data provided by instant police reporting applications brings not only a quantitative change (which might be rather small) but also a qualitative change: the selection of the data added could be skewed towards certain groups, as was pointed out in the previous subsection. Furthermore, the choice of who to add to the database is filtered by citizens on the lookout for transgressive behaviors. This might frame everyone captured in the submitted image as worthy of at least a cursory investigation, if not as a suspect then at least as a witness. Thus, having one's data included in surveillance databases as a result of instant police reporting heightens the chance of losing your anonymity vis-à-vis the police now or at a later point in time.

Normalization and chilling effects

An important effect of the possible addition of (potentially) identifiable data to surveillance databases as a result of instant police reporting applications is normalization. This subsection investigates what the substantive normalizing effects might be using post-panoptical theories.

The strict norms of the frequent users of instant police reporting applications might mean that acts which are illegal, but of which it is generally accepted that they go unpunished most of the time, will be reported to the authorities. Although there are probably not enough police officers to follow up on each minor infraction, this does not suggest that abundant reporting will be without consequences altogether. Primarily, data about offenders may be added to the local police database of "known offenders" that is only acted upon at a later point in time – for instance, if the number of transgressions by the same person becomes excessive. Furthermore, institutions faced with limited resources might wish to automate the process of handing out fines for minor infractions using biometrics, in a manner similar to how many speeding tickets are currently issued using automatic number plate recognition (ANPR) software. Alan Westin (1967, 35) predicted that such unfettering scrutiny would mean that "most persons in society would be [...] in jail." Although this might be overly pessimistic, the above does imply a normalizing pressure to not engage in petty illegal behavior whenever others are present.

Because they engage in acts that are designed to attract substantial attention, protestors form an especially interesting category with regard to normalization. In order for them to succeed, they need to tiptoe around, and sometimes cross over, the edges of social, moral, and legal norms and do so in places where they are highly visible; an open invitation to instant police reporting users. Chilling effects on free speech occur when protestors are forced to consider whether their cause is more important to them than their existence (or lack thereof) in a police database: will I protest the new multi-storey parking garage, or will I remain anonymous to the police? Note that protestors have always had to consider such questions and that it might not influence the most determined amongst them. However, instant police reporting applications can also dramatically increase the visibility of more casual protestors, which increases their chances of being included in a police database (see also Schneider and Trottier 2012).

The erosion of the presumption of innocence and social trust

Threats to the presumption of innocence and social trust arise when instant police reporting applications are construed as measures of mass surveillance. Such a construction may seem counter-intuitive at first: any recordings made are obviously targeted towards specific transgressive behaviors and those suspected of committing them. However, the ubiquity of smartphones means that the minutiae of everyday life can be captured and shared with the police at any moment. Coupled with the possibilities for virtually endless retention of data and the inherent biases of users, this means that for citizens (and especially those citizens in already marginalized positions) the effects of instant police reporting applications can be analogous to those of other mass surveillance measures: anyone can at any time be added to a database for the sole purpose of using the gathered data later during a possible investigation. Such a pre-emptive logic presents a threat to the principle of the presumption of innocence: data is gathered by the police without a specific suspicion, which shifts the burden of proof and dilutes the right to remain silent (Milaj and Mifsud Bonnici 2014).

Another effect of measures of mass surveillance is the corrosion of social trust in society. The co-optation of citizens in mass surveillance efforts, in particular, eats away at such interpersonal trust: citizens are asked to approach one another as possible suspects, and must take into account that they may be approached in this way by others as well (Marx 2013). Maria Los (2006) relates such strategies explicitly to those employed by the internal security services of former communist states in Eastern Europe and points out the atomization of society which results: "[E]ach individual not only views all others as potential spies but must also be aware of being similarly viewed by others. This creates painful barriers of fear and humiliation that divide and terrorize society" (2006, 83). One only needs to imagine what might happen if, during a wild night out in the city, someone in a group of friends takes out her/his phone to record a video. Timan and Oudshoorn (2012) found that, in such situations, the primary concern of those documented was the dissemination of the material; a concern which is currently mostly connected to the lack of control over privacy on social networking sites (Fox and Moreland 2015, 171–72). Add the police as a potential recipient and, in a society saturated with smartphones, such situations are apt to spiral out of control and lead to a loss of generalized societal trust.

Cumulative disadvantages resulting from simulation and prediction

All the effects pointed out so far in this section are most pronounced on those people whose data is collected in surveillance databases. As discussed above, their makeup is not neutral. This is deeply troublesome, as it means that citizens in already marginalized positions will be especially affected by the rhizomatic coupling of databases. This phenomenon is known as a cumulative disadvantage: a way in which past negative decisions tend to cluster future hardships on certain persons or groups (Gandy 2006, 319; Gandy 2000, 1100–01).

Related to and overlapping with cumulative disadvantages are the final and perhaps most nebulous effects of the accumulation of identifiable and potentially identifiable data as a result of instant police reporting applications: the effects related to simulation and prediction. A photograph taken with a smartphone is an altogether different beast than a Polaroid, and mostly so because of the metadata embedded in the photo itself or in the message used to send the photo. These include time and location which might be important for the purposes of prediction; they are two of the three data points used by PredPol,[13] one of the most widespread commercial software packages which police forces can use to guestimate when and where crimes might take place and thus where police presence is needed.

However, inputting data obtained from instant police reporting applications into such predictive algorithms runs the risk of reifying user biases. Such software influences the surveillance discretion of the police, e.g., when and where they will be looking for crime (Joh 2016). If users consistently send in reports from certain areas, the police will, given enough prods by the algorithm, inevitably find crime there. This retroactively justifies the many reports, and will in turn lead to even more patrolling and more biased data being fed into the algorithms.[14] As a result, the individuals and communities living in neighborhoods where a relatively high number of reports are being filed will suffer from a further amplification of the surveillance tax – the chilling effects and negative social stigma that results from increased crime and police scrutiny (Joh 2016, 32; Mantelero 2016, 240).

Policy implications

This chapter has a clear focus towards the future: instant police reporting applications (at the time of writing) are not yet a widespread phenomenon. Throughout the theoretical body of this chapter, formed by the previous two sections, various strands of ideas and arguments surfaced which may critically shape the future of citizen involvement in crime fighting using smartphones. Three implications follow from this chapter for policy makers who wish to gain from instant police reporting applications the highest possible reduction in criminality combined with the highest level of human rights protection possible.

A first implication, which follows from the panoptical diagram, is that the widespread uptake of instant police reporting applications will be critical in determining whether possible transgressors will modify their behavior. Lobbying standard-setting organizations to grant instant police reporting functionalities the same status as emergency calls seems to be the most effective option here. Another important actor in the promotion of instant police reporting applications might be the mass media in the broadest sense possible. They provide the narratives through which people make sense of surveillance (Kammerer 2012, 99) and, as such, can directly and indirectly impact potential transgressors. If potential transgressors are confronted with stories in which a criminal has been caught on the basis of smartphone reporting – and attach some value to these stories (Doyle 2011) – they might decide not to proceed. The media can also impact possible transgressors indirectly, by presenting smartphone apps for instant police reporting to the general public as an essential tool for personal safety and thus contributing to their widespread uptake.

Second, many of the negative effects outlined in the section entitled "Societies of control" are a result of overzealous reporting. These effects can be dampened if the police employs its discretionary power once a report has been filed not just to decide if officers on the ground are dispatched, but also to decide how the personal data from the report will be handled.[15] Operators receiving reports (whether emergency call-takers or specialized operators) should select individuals from images for addition to databases, rather than outsourcing this process to an algorithm that might add anyone. Although more costly in the short term, this can ensure the privacy of bystanders and somewhat dampen the cumulative disadvantage which results from the discriminatory biases of app users. In the long run, this will heighten the acceptance of instant police reporting applications, resulting in more and more valuable usage of the apps. Here, the role of the media must again be emphasized: instant police reporting applications should be promoted as tools only to be employed in life-or-death situations similar to emergency telephone numbers in order to prevent overzealous reporting.

Finally and most importantly, continued and careful attention to vulnerable groups is warranted throughout the operation of instant police reporting applications. Especially the re-use of the data obtained through instant police reporting applications should be approached with extreme caution. In the section above, only one potential effect of the use of data from instant police reporting applications for the purposes of prediction has been discussed, but the possibilities for re-use are manifold. However, any dataset (even partly) based on instant police reporting should be considered biased from the outset, which invariably taints the results from any subsequent analyses. Basing policies on such results is ill-advised, and risks instituting discriminatory procedures.

Conclusion

This chapter set out to investigate the societal effects of instant police reporting applications for smartphones using theories from surveillance studies. The chapter was split in two following Deleuze's (1992, 3) claim that the societies of discipline as described by Foucault have been superseded by societies of control. His analysis is often taken to mean that any theoretical study of contemporary surveillance using panoptical schemas is essentially flawed. However, this may be too simplistic (Simon 2005, 2): this chapter has shown that the combination of both panoptic and post-panoptic theoretical frameworks can, if applied critically and carefully, enhance and broaden our understanding of a surveillance phenomenon.

At the time of writing, instant police reporting applications are an emerging phenomenon. It might be that instant police reporting applications will lead a life lingering in the depths of various app repositories to be installed from time to time but rarely used. However, that law enforcement agencies will have increasing amounts of visual recordings of suspicious behaviors made available to them – whether through dedicated platforms, social networking sites, or e-mail – by the general public and that the possibilities for analysis and storage of such materials are growing seems evident. The results from this chapter therefore have broader relevance.

Instant police reporting applications request that we trust fellow citizens to judge whose anonymity may be sacrificed to facilitate the work of the police – holders of the exclusive right to intra-territorial violence in order to maintain social order. The section "Societies of discipline" has demonstrated the power that potential users of such apps wield. However, with great power there must also come great responsibility (Lee 1962). The extent of this responsibility was established in the section "Societies of control" where the societal effects of the loss of anonymity were discussed: pressures towards conformity, the chilling of free speech, the erosion of social trust and the presumption of innocence, and how these might disproportionally affect already disadvantaged individuals and groups. An essential question, which can and should not be answered in an academic publication, but in the public debate, thus remains: how great is the power and responsibility that we are willing to grant the public in supporting the police in the execution of their tasks?

Avenues for future research

There are certain questions which do belong in the academic debate. This chapter has sketched out what the societal effects of instant police reporting applications might be in the very near future, and there is abundant room for more in-depth research. This subsection lists, non-exhaustively, various avenues for future research in multiple academic fields, but much interdisciplinary work also remains. Independent of the field of research, a special sensitivity towards possible discriminatory tendencies is warranted.

What is needed first and foremost is continuing scrutiny to document how instant police reporting applications, or more generally, the ways in which citizens make visual recordings available to the police, will advance in the coming years. Especially interesting will be a Science and Technology Studies approach focused on specific implementations to study how the developments described in this chapter will play out on the streets, and how the user base of such apps will develop. The sensitivity of Science and Technology Studies to non-use (Dubbeld 2006, 192) is also interesting from a policy perspective: what are prohibiting and facilitating factors for citizens and police officers in the context of the use of instant police reporting applications? Will users engage in acts of trolling, for instance by flooding the system with meaningless pictures, and how will police forces respond to such behavior?

Related to this question is the perhaps more criminological concern: are there certain transgressions which lend themselves especially well to reporting using smartphones? If there are, how does that impact the execution of police tasks and the distribution of police resources? Think here for instance of the obvious discrepancies in visibility between public intoxication and white collar fraud: will differences in visibility between various crimes shift police resources in ways which favor or disfavor certain groups? And could citizens turn this process on its head to give more visibility to crimes they deem important, for instance through collective actions?

Another topic which will merit assiduous dissection in the years to come is the role of the private sector in the development and ongoing support of instant police reporting applications. This chapter has focused on the relationship between the police and citizens as their respective roles are, to a certain degree, foreseeable. However, any part of the analysis presented here could be deeply influenced by the involvement of the private sector, which is already taking a lead in related technologies. For example, Vigilant Solutions (2014) maintains a database of over 2 billion ANPR readings, and Taser International provides a hosting solution for the videos made with their popular Axon police body-worn cameras (Gelles 2016). It thus seems likely that these or similar companies will also play an important role in instant police reporting applications. However, this raises profound questions: what are the judicial and democratic controls on them? What will the bargaining position of states and local police departments be, both during initial negotiations and once a system has been deployed? Furthermore, do companies stand to gain from incentivizing use, financially or otherwise, and how will this influence the effects instant police reporting applications have on society? Thorough analysis using the surveillant assemblage theoretical framework seems especially suited for investigation of these highly relevant and under-researched questions.

This chapter has analyzed possible societal consequences of instant police reporting apps using theories on surveillance and social control. Although perhaps only a small shift, an analysis focused on privacy rather than surveillance might highlight how instant police reporting apps impact other processes including, for instance, identity-building (Cohen 2013). Such an analysis is especially salient with regard to teenagers who engage in complex privacy behaviors vis-à-vis authority figures and institutions (boyd and Marwick 2011) but who are largely dependent on public spaces to escape the gaze of parents and educational institutions.[16] They may feel that their spaces for identity formation are being threatened by technophiliac police forces in cahoots with overzealous citizens, but more considerate scrutiny is needed. Attention to the rhizomatic tying of surveillance assemblages is needed in this context as such linkages may lead to "context collapse", in which the carefully separated private spaces of teenagers (boyd and Marwick 2011) start meshing together. The role that social networking and sharing platforms play in the lives of teenagers and young adults deserves special attention in this regard (e.g. Schneider and Trottier 2012; Timan and Albrechtslund 2015).

Instant police reporting applications also offer an impetus for further research by cultural geographers who address the role of popular mobile communication and computation technologies in the experience of spaces.[17] Their perspective on the implications of the active involvement of state actors (and specifically those state actors with a claim to force) in such intimate and quotidian technologies could enhance our understandings of contemporary cities, saturated as they are with smartphones and security concerns. Particularly interesting might be a study of how the erosion of social trust is spatially distributed: is it more prevalent in, e.g., public transport, where "see something, say something" campaigns are already widespread, or are such effects larger in spaces which are traditionally less associated with security threats?

Finally, legal scholars might consider the impact of instant police reporting applications on data protection frameworks. The meshing of citizen and police roles problematizes the often strict separation between data collecting and processing regimes for private persons and law enforcement agencies. Concepts such as notification and consent, which are central to data protection frameworks for non-governmental purposes around the globe, seem inept to deal with the dual pressures of technology and security (Haggerty 2015): should users of instant police reporting applications literally shout across the street to inquire if criminals consent to being filmed?

Acknowledgments and author's note

I gratefully acknowledge the many helpful comments received from an anonymous reviewer and the editorial team which helped to strengthen the argument and provided various avenues for connecting the chapter to broader debates and literature. Many thanks are also due for the constructive feedback I received on early drafts of this chapter by Prof. Mifsud Bonnici. Early ideas for this chapter were presented in November 2015 at the Cyberspace '15 conference of the Masaryk University in Brno, Czech Republic, where I received several helpful comments.

The author is pursuing a PhD under the supervision of Prof. Mifsud Bonnici of the University of Groningen, who also led the Horizon2020 CITYCoP (Citizen Interaction Technologies Yield Community Policing) project, funded by the European Commission under grant agreement no. 653811. One of the main results of CITYCoP is a smartphone application aimed at increasing community policing efforts which includes limited instant police reporting possibilities. The views expressed in this chapter are those of the author alone and are in no way intended to reflect those of the European Commission and/or the CITYCoP consortium.

Notes

1 See for a typical example of the predominance of voice calls for emergency situations the Dutch Politie app (the official app of the Dutch police) which redirects the user to calling 112 when s/he tries to report an emergency (Nationale Politie 2015).
2 For related projects see also (Community Research and Development Information Service 2016b; Community Research and Development Information Service 2016c; Community Research and Development Information Service 2016d).
3 For an artist's rendition of such possibilities see for instance Nolan's *The Dark Knight* (2008) which features a botnet which turns all smartphones in the fictional Gotham City into echolocation devices.
4 Note that the term "image" is used in this chapter to denote any still and/or a moving image material, i.e. both photo and video materials are covered by this term.
5 The term "CCTV system" is used in this chapter as a shorthand for any security camera system in a public space, as is common usage. The term is not meant to denote whether or not the circuits that the security cameras are attached to are closed or (as is increasingly prevalent) interconnected to other systems.
6 Based on information provided by the reporter or location data appended to either the image or the report.
7 It bears mention here that CCTV systems are often used by the police for myriad other purposes as well, such as providing a helicopter view of a developing situation and providing intel to officers on the scene. Such purposes are outside the scope of this chapter.
8 Although no legal requirements exist in this regard, the standardization group ETSI requires emergency calls to always be possible from a mobile phone (except when it is turned off) since at least 1998, making it a de facto requirement on phone manufacturers (ETSI 2015, 26).
9 Among the more intriguing examples in this category are people who fashion themselves superhero personas to fight crime. For many of them, reporting to the police is an important part of being a real-life superhero (Fishwick and Mak 2015, 347).
10 These studies were selected not because training might have been a panacea for the problems presented, but because they offer examples that show some similarity with instant police reporting applications, insofar that their users will also be more-or-less untrained.
11 Since Figure 9.1 is only presented here for illustrative purposes, it should be mentioned that the central figure is probably also quite innocuous.
12 In *Discipline and Punish* Foucault (1995, 277–78) considered this an essential moment as it signals the move from illegality, which may be tolerated, towards delinquency, which should be corrected and normalized. Instant police reporting applications put this power in the hands of citizens.
13 The final data point being the type of crime (PredPol 2015).
14 Such loops might very well negate the possible advantage which predictive algorithms present for creating a more equitable distribution of surveillance discretion (Joh 2007; Joh 2014, 58).
15 Compare here also note 12.
16 See Carissa Véliz's contribution in this volume for a wider consideration of those who rely on public space for privacy.
17 For a more general call for increased attention to popular mobile technologies in cultural geography see Kinsley (2016).

References

boyd, danah, and Alice E. Marwick. 2011. "Social Privacy in Networked Publics: Teens' Attitudes, Practices, and Strategies." In *A Decade in Internet Time: Symposium on the Dynamics of the Internet and Society*. Accessed March 23, 2016. https://papers.ssrn.com/sol3/papers.cfm?abstract_id=1925128.

Certeau, Michel de. 1984. *The Practice of Everyday Life*. Berkeley: University of California Press.

Cohen, Julie E. 2013. "What Privacy is for." *Harvard Law Review* 126: 1904–1933.

Community Research and Development Information Service. 2015a. "Citizen Interaction Technologies Yield Community Policing (CITYCoP)." *CORDIS*. Accessed November 12. 2015. http://cordis.europa.eu/project/rcn/197273_en.html.

Community Research and Development Information Service. 2016b. "Inspiring CitizeNS Participation for Enhanced Community PoliCing AcTions (INSPEC2T)." *CORDIS*. Accessed May 18, 2016. http://cordis.europa.eu/project/rcn/194895_en.html.

Community Research and Development Information Service. 2016c. "TRusted, CItizen – LEA colLaboratIon over sOcial Networks (TRILLION)." *CORDIS*. Accessed May 18, 2016. http://cordis.europa.eu/project/rcn/194841_en.html.

Community Research and Development Information Service. 2016d. "Unity." *CORDIS*. Accessed May 18, 2016. http://cordis.europa.eu/project/rcn/194893_en.html.

Deleuze, Gilles. 1992. "Postscript on the Societies of Control." *October* 59 (Winter): 3–7.

Doyle, Aaron. 2011. "Revisiting the Synopticon: Reconsidering Mathiesen's 'The Viewer Society' in the Age of Web 2.0." *Theoretical Criminology* 15(3): 283–299.

Dubbeld, Lynsey. 2006. "Telemonitoring of Cardiac Patients: User-Centred Research as Input for Surveillance Theories." In *Theorizing Surveillance: The Panopticon and Beyond*, edited by David Lyon, 182–205. Abingdon: Routledge.

ETSI. 2015. "Universal Mobile Telecommunications System (UMTS); LTE; Service Aspects; Service Principles (3GPP TS 22.101 Version 11.10.0 Release 11)." Accessed January 26, 2016. www.etsi.org/deliver/etsi_ts%5C122100_122199%5C122101%5C11.10.00_60%5Cts_122101v111000p.pdf.

Federal Bureau of Investigation. 2016. "Next Generation Identification." FBI. Accessed July 15, 2016. www.fbi.gov/about-us/cjis/fingerprints_biometrics/ngi.

Fishwick, Elaine, and Heusen Mak. 2015. "Fighting Crime, Battling Injustice: The World of Real-Life Superheroes." *Crime, Media, Culture* 11(3): 335–356. doi:10.1177/1741659015596110.

Foucault, Michel. 1995. *Discipline and Punish: The Birth of the Prison*. Translated by Alan Sheridan. 2nd Vintage Books ed. New York: Vintage Books.

Fox, Jesse, and Jennifer J. Moreland. 2015. "The Dark Side of Social Networking Sites: An Exploration of the Relational and Psychological Stressors Associated with Facebook Use and Affordances." *Computers in Human Behavior* 45 (April): 168–176. doi:10.1016/j.chb.2014.11.083.

Galič, Maša, Tjerk Timan, and Bert-Jaap Koops. 2016. "Bentham, Deleuze and Beyond: An Overview of Surveillance Theories from the Panopticon to Participation." *Philosophy & Technology*, May. doi:10.1007/s13347-016-0219-1.

Gandy, Oscar H. Jr. 2000. "Exploring Identity and Identification in Cyberspace." *Notre Dame Journal of Law, Ethics & Public Policy* 14(2): 1085–1111.

Gandy, Oscar H. Jr. 2006. "Quixotics Unite! Engaging the Pragmatists on Rational Discrimination." In *Theorizing Surveillance: The Panopticon and Beyond*, edited by David Lyon, 318–336. Abingdon: Routledge.

Gelles, David. 2016. "Taser International Dominates the Police Body Camera Market." *The New York Times*. July 12. Accessed December 15, 2016. www.nytimes.com/2016/07/13/business/taser-international-dominates-the-police-body-camera-market.html.

Goff, Tanguy le, Virginie Malochet, and Tiphaine Jagu. 2011. "Surveiller à distance. une ethnographie des opérateurs municipaux de vidéosurveillance." Île-de-France: Institut d'aménagement et d'urbanisme.

Goold, Benjamin J. 2006. "Open to All? Regulating Open Street CCTV and the Case for Symmetrical Surveillance." *Criminal Justice Ethics* 25: 3–17.

Groombridge, Nic. 2002. "Crime Control or Crime Culture TV?" *Surveillance & Society* 1(1): 30–46.

Haggerty, Kevin D. 2015. "What's Wrong with Privacy Protections? Provocations from a Fifth Columnist." In *A World without Privacy: What Law Can and Should Do*, edited by Austin Sarat. New York: Cambridge University Press.

Haggerty, Kevin D., and Richard V. Ericson. 2000. "The Surveillant Assemblage." *British Journal of Sociology* 51(4): 605–622.

Harshaw, Pendarvis. 2015. "Nextdoor, the Social Network for Neighbors, is Becoming a Home for Racial Profiling." *Fusion*. March 24. Accessed December 12, 2016. www.fusion.net/story/106341/nextdoor-the-social-network-for-neighbors-is-becoming-a-home-for-racial-profiling/.

Hier, Sean P., Kevin Walby, and Josh Greenberg. 2006. "Supplementing the Panoptic Paradigm: Surveillance, Moral Governance and CCTV." In *Theorizing Surveillance: The Panopticon and Beyond*, edited by David Lyon, 230–244. Abingdon: Routledge.

Joh, Elizabeth E. 2016. "The New Surveillance Discretion: Automated Suspicion, Big Data, and Policing." *Harvard Law & Policy Review* 10(1): 15–42.

Kammerer, Dietmar. 2012. "Surveillance in Literature, Film and Television." In *Routledge Handbook of Surveillance Studies*, edited by Kirstie Ball, Kevin D. Haggerty, and David Lyon, 99–106. Abingdon: Routledge.

Kinsley, Samuel. 2016. "Vulgar Geographies? Popular Cultural Geographies and Technology." *Social & Cultural Geography* 17(6): 793–796. doi:10.1080/14649365.2016.1152394.

Koskela, Hille. 2003. "'Cam Era' – the Contemporary Urban Panopticon." *Surveillance & Society* 1(3): 292–313.

Lauer, Josh. 2012. "Surveillance History and the History of New Media: An Evidential Paradigm." *New Media & Society* 14(4): 566–582. doi:10.1177/1461444811420986.

Lee, Stan. 1962. "Spider-Man!" In vol. 15 of *Amazing Fantasy*. New York: Marvel Comics.

Leistert, Oliver. 2013. *From Protest to Surveillance: The Political Rationality of Mobile Media: Modalities of Neoliberalism*. Frankfurt am Main; New York: PL Academic Research.

Levin, Sam. 2016. "What Happens When Tech Firms End up at the Center of Racism Scandals?" *The Guardian*. September 30. Accessed December 12, 2016. www.theguardian.com/technology/2016/aug/30/tech-companies-racial-discrimination-nextdoor-airbnb.

Los, Maria. 2006. "Looking into the Future: Surveillance, Globalization and the Totalitarian Potential." In *Theorizing Surveillance: The Panopticon and Beyond*, edited by David Lyon, 69–94. Abingdon: Routledge.

Mantelero, Alessandro. 2016. "Personal Data for Decisional Purposes in the Age of Analytics: From an Individual to a Collective Dimension of Data Protection." *Computer Law & Security Review* 32: 238–255.

Marx, Gary T. 2013. "The Public as Partner? Technology Can Make Us Auxiliaries as Well as Vigilantes." *IEEE Security & Privacy* 11(5): 56–61. doi:10.1109/MSP.2013.126.

Milaj, Jonida, and Jeanne Pia Mifsud Bonnici. 2014. "Unwitting Subjects of Surveillance and the Presumption of Innocence." *Computer Law & Security Review* 30(4): 419–428. doi:10.1016/j.clsr.2014.05.009.

Nationale Politie. 2015. "Politie App." Accessed December 3, 2015. www.politie.nl/app.

Nederlandse Omroep Stichting. 2016. "Contact Met de Politie? Stuur Een Appje!" *NOS*. December 8, 2016. http://nos.nl/artikel/2147234-contact-met-de-politie-stuur-een-appje.html.

Nolan, Christopher. 2008. *The Dark Knight*. Burbank, C: Warner Bros.

Norris, Clive, and Gary Armstrong. 1999. "CCTV and the Social Structuring of Surveillance." *Crime Prevention Studies* 10: 157–178.

PredPol. 2015. "How PredPol Works: We Provide Guidance on Where and When to Patrol." PredPol. Accessed May 9, 2016. www.predpol.com/how-predpol-works/.

RapidSOS. 2016. "Haven." Accessed December 9, 2016. http://www.rapidsos.com.

Reilly, Paul. 2015. "Every Little Helps? YouTube, Sousveillance and the 'Anti-Tesco' Riot in Stokes Croft." *New Media & Society* 17(5): 755–771. doi:10.1177/1461444813512195.

Schneider, Christopher J., and Daniel Trottier. 2012. "The 2011 Vancouver Riot and the Role of Facebook in Crowd-sourced Policing." *BC Studies* (175): 57–72.

Simon, Bart. 2005. "The Return of Panopticism: Supervision, Subjection and the New Surveillance." *Surveillance & Society* 3(1): 1–20.

Smith, Gavin J. D. 2016. "Surveillance, Data and Embodiment: On the Work of Being Watched." *Body & Society* 22(2): 108–139. doi:10.1177/1357034X15623622.

The National 911 Program. 2016. "The Future of 911." *911.gov*. Accessed December 19, 2016. www.911.gov/futureof911.html.

Timan, Tjerk, and Anders Albrechtslund. 2015. "Surveillance, Self and Smartphones: Tracking Practices in the Nightlife." *Science and Engineering Ethics*, August. doi:10.1007/s11948-015-9691-8.

Timan, Tjerk, and Nelly Oudshoorn. 2012. "Mobile Cameras as New Technologies of Surveillance? How Citizens Experience the Use of Mobile Cameras in Public Nightscape." *Surveillance & Society* 10(2): 167–181.

Trottier, Daniel. 2014. "Crowdsourcing CCTV Surveillance on the Internet." *Information, Communication & Society* 17(5): 609–626. doi:10.1080/1369118X.2013.808359.

Vigilant Solutions. 2014. "LPR Data." Accessed December 15, 2016. www.vigilantsolutions.com/wp-content/uploads/2014/08/LPR_Data_AUG2014_LFS.pdf.

Welsh, Brandon C., and David P. Farrington. 2002. "Crime Prevention Effects of Closed Circuit Television: A Systematic Review." Home Office Research Study 252. London: Home Office Research, Development and Statistics Directorate. Accessed May 20, 2016. http://webarchive.nationalarchives.gov.uk/20090410070401/http://www.homeoffice.gov.uk/rds/pdfs2/hors252.pdf.

Westin, Alan F. 1967. *Privacy and Freedom*. London: The Bodley Head.

Winston, Ali. 2015. "Los Angeles Sheriff Invests in New Tech to Expand Biometric Database." *Reveal.* July 3. Accessed April 13, 2016. www.revealnews.org/article/los-angeles-sheriff-invests-in-new-tech-to-expand-biometric-database/.

10
"I'm a creep, I'm a weirdo"[1]
Street photography in the service of the male gaze

Stuart Hargreaves

Introduction

Freud described the sexual pleasure gained from looking upon another as scopophilia ("the love of looking"). Mulvey famously argued that conventional cinematic structures seek to replicate this sexual pleasure, because "although the film is really being shown … conditions of screening and narrative conventions give the spectator the illusion of voyeuristic separation" (Mulvey 1975, 9). Importantly, though, the particular form these structures took represented a controlling "male gaze" with women as its object.

> In a world ordered by sexual imbalance, pleasure in looking has been split between active/male and passive/female. The determining male gaze projects its phantasy on the female figure which is styled accordingly.
> (Mulvey 1975, 11)

Some 40 years after Mulvey's work, we can also readily find the male gaze in new digital forms of art (Schleiner 2001; Adam 2002; Sarkeesian 2016). This chapter considers two new forms of this gaze, both of which involve the digitally mediated collection and sharing of photographs of women in public, without their knowledge or consent. The first is the phenomenon of "creepshots," in which individuals surreptitiously take photographs of women and subsequently upload them to message boards for evaluation and commentary by other users. The second involves individuals scouring virtual street maps (such as Google Street View) for images of women they find "notable" in some way. Once discovered, these images are "cut" and placed online for the purposes of evaluation and commentary. In both cases, the subjects of the images are in public or quasi-public spaces. In both cases, women are the overwhelming targets of this digital gaze. In both cases, the internet mediates the dissemination of the images, creating the potential for a large and anonymous audience. In both cases, the online commentary quickly becomes sexual in nature, and is frequently overtly hostile.

Though perceiving this gaze as a series of invasions of the subject's privacy is understandable, it is ultimately unhelpful for at least two reasons. First, actionable privacy claims can generally only be made in a limited range of circumstances, into which it is exceedingly difficult to fit this kind of image gathering and sharing. In fact, the inability of privacy laws to readily deal with this issue is reflected in the fact that the participants often rely on traditional legal accounts of privacy to *defend* their actions, arguing that individuals have no reasonable expectation of privacy while in public. Second, even if individual legal remedies *could* be obtained, this would miss the more significant harm: the acceptance or reinforcement of the notion that women in public exist as sexual objects for the interest of others. Rather than being a problem of privacy, then, creepshots and the third-party use of virtual street-map images discussed in this chapter are better understood as new forms of public surveillance, with a particularly gendered bent. As surveillance practices, they implicate different kinds of harms – broader, more diffuse – than do conventional privacy invasions, so solutions cannot be found simply by attempting to provide legal remedies for those who find themselves in an image that is being circulated online without their consent.

"Creepshots"

"Creepshots" – a descriptor willingly adopted and embraced by participants – refers to the practice of surreptitiously taking photos of women in public places (parks, beaches, streets, etc.) or publicly accessible places (schools, retail outlets, gyms, etc.), then posting them online for sexualized evaluation, commentary, and discussion. Though the descriptor may be tacit acknowledgment by the participants of their status as "creeps", it seems more likely to stem from the idea that surreptitiously taking a photo may require "creeping up" on the subject. The phenomena appears to have originated on the well-known website reddit.com, which bills itself as "the front page of the internet." It features "subreddits" created and run by volunteers on virtually every subject under the sun – politics, culture, computers, sports, relationships, education, movies, art, etc. "Redditors" (users of reddit) have historically adopted a *laissez-faire* approach to content with effectively no limits on what topics are discussed or in what manner. This openness means reddit is filled with knowledgeable commentators and interesting discussions; President Obama, for instance, once hosted a live session in the "Ask Me Anything" subreddit, in which users state their job and then answer questions about it posed to them by other users (Reddit 2012a). However, the general lack of oversight that characterizes most of reddit's history *also* means that there are rather darker areas as well (Warzel 2015). One of these areas was the creepshots subreddit (hereafter "r/creepshots", in the parlance of redditors).

R/creepshots was set up by an anonymous redditor who encouraged users to "spread the word and get creepin [sic]" (Baker 2012a). Members of r/

creepshots duly obliged and uploaded photos of friends, co-workers, fellow students, and random strangers for sexual judgment and discussion; hundreds of photographs were submitted. To take an emblematic example, redditor "RandomGuy67B" described a submitted photograph of a girl wearing a short skirt and a religious necklace as a "great post", leading "nerdyfitnessguy" to agree and rhetorically ask "who *doesn't* want to pound a church slut?" (Imgur n.d.a). The style of the photos – from behind the subject, or with the subject looking away, or with the use of telephoto lens – suggested that the subjects were unaware that they were being photographed. The women were not able to give or deny consent to being photographed, because they were unaware it was even happening. I will show later in this chapter that not only is a lack of consent on the part of the subject fundamental to the nature of a "creepshot", participants argue that it is irrelevant thanks to a legal principle that suggests individuals in public do not have a "reasonable expectation of privacy".

Eventually, r/creepshots broke through into the larger reddit consciousness thanks to its growing popularity (reddit "promotes" popular subreddits and message threads to its front-page, where all redditors can see them). This led to redditors who were not members of r/creepshots seeing the material, and subsequent complaints that many of the images appeared to show secondary school students – minors. Redditor "SpencerTracyMorgan" highlighted the apparent encouragement of predatory behavior in schools, citing the praise received by one uploader from another member of r/creepshots:

> [I admire] the dedication you put into [this series of photos]. You followed this girl down a long hallway and beyond and still managed to center each shot, even though they are lopsided. You maintained a distance of a few feet the whole time even though there were few people going your direction.
>
> (Reddit 2012b)

Disturbed, "SpencerTracyMorgan" led a crowdsourced investigation of a number of photos captioned as "hot senior chick in one of my classes" that appeared to be taken from the vantage point of the front of a classroom (Reddit 2012b). Redditors eventually determined these images had been taken by a substitute high school teacher; he was ultimately fired (and later arrested when a search of his phone revealed sexually explicit text and photo exchanges with minors) (Nelms 2012). Awareness of r/creepshots gradually filtered out across the internet and though complaints grew, reddit's owners initially declined to ban the subreddit, arguing it did not break any content rules (Herman 2012). In response, one redditor took matters into her own hands by starting a blog called "Predditors" (Predditors 2012) in which she attempted to "doxx" redditors participating in r/creepshots (an attempt to try and connect their anonymous online identity with their real-world identity) (Baker 2012b). When the moderator of r/creepshots was confronted with the possibility

(Imgur n.d.b) of being publicly "outed" in this way, they first made r/creepshots "private" (meaning only registered members could view the images) and then shut it down entirely (Reddit 2012c). A game of "whac-a-mole" between the site's management and a minority of its users ensued, with a series of similar subreddits with names like r/creepsquad and r/creepyshots appearing, only to then be shut down as part of the ongoing clean-up (Hern 2015).

Yet, several years later, creepshots-style material can still be readily found on reddit. Frequently, subreddits seek to skirt the rules by pretending to be something they are not. Take for instance r/candidfashionpolice, which survived for close to three years following the initial cull of creepshots-style subreddits by attempting to spin the purpose of the images uploaded by users as something non-sexual. Featuring stylistically identical content to r/creepshots, the users of r/candidfashionpolice nonetheless pretended that the photos were taken purely out of an interest in fashion, with users placing comments on surreptitiously taken photos such as "I don't think those leggings go with that bag." Their inability to stay entirely on-message, however, led to r/candidfashionpolice being finally banned in late 2015; at the time, it had 66,000 members (Reddit 2015). Other subreddits with similar content continue to survive, however. R/volleyballgirls describes itself as devoted to "The glory of volleyball," though that phrase is in fact found within a longer sentence with certain words struck out: "The glory of ~~the officially-sanctioned~~ volleyball ~~outfit~~," a winking acknowledgment of its true purpose (Reddit n.d.a). It contains a collection of photographs of female high school and university students standing on volleyball courts, usually taken from behind. The content is presented in precisely the same manner as that found on r/creepshots – for instance, a photo uploaded at the time of writing was titled "Down to her ass," an obvious invitation to sexualized commentary from other members of r/volleyballgirls that was readily accepted (Reddit 2016). In a similar vein, r/yogapants explicitly directs its users to "not submit creep shots" in an apparent attempt to remain in compliance with reddit's new policies (Reddit n.d.b). But much like the "mission statement" of r/volleyballgirls, the publicly stated goal of compliance with reddit's new creepshots policy by the moderators of r/yogapants appears to be inconsistent with its actual operation. Its moderators make no effort to enforce that rule, and so r/yogapants features many photos of women that appear to have been taken without their knowledge, along with the expected sexualized commentary. At the time of writing both subreddits remain publicly accessible and feature new creepshots-esque content uploaded by users on a regular basis.

Of course, even if reddit's management were able to successfully purge all relevant material from their servers, it can simply reappear elsewhere. After r/candidfashionpolice was shut down on reddit, for instance, it returned (and remains accessible at the time of writing) as v/Candid on voat.co, a reddit-style message board that attempts to recreate the "no rules" aspect of reddit's early days (voat n.d.). Though voat.co does not explicitly ban creepshots material, the moderators of v/Candid nonetheless continue to attempt to maintain a façade of fashion critique, including posting a series of "rules" including "titles and comments must pertain to fashion" and "no heterosexual men [allowed]. We strive to make this a female and queer friendly place to discuss terrible fashion" (voat n.d.). Again, this kind of mission statement appears to be nothing more than a ploy to divert scrutiny; spend any time in the forum and its true purpose quickly becomes apparent. Indeed, even some its members appear to be tiring of maintaining the façade: after uploading a video of a young woman in tight yoga leggings walking around a supermarket (surreptitiously recorded by someone following her without her knowledge), a user titled it "Do I have to pretend this is about fashion anymore?" (voat 2015a). As was the case on reddit, the comments on voat quickly go beyond the juvenile. In an emblematic example, after "reviewing" a surreptitiously taken photo of a young woman in a revealing outfit on v/Candid, one user wrote, "God she is absolutely the reason bad things happen to girls on the subway" (voat 2015b). On this account, not only are women objects that exist in public for the titillation and arousal of men, but should they prove too attractive then they are also responsible for "enticing" men to physically assault them.

The phenomenon of creepshots extends beyond reddit and its clones. There is at least one entirely dedicated website (http://creepshots.com) and an associated twitter account (http://twitter.com/creepbj/) for "fans" to follow. In contrast to its reddit progenitor, this website has "gone commercial" (though seemingly still reliant in part on its own users for its content): it charges users USD 1.33/month and promises "over 7000 candid voyeur creepshots in 60 different categories," with "new creepshots added daily; new video creeps added every week" (CreepShots n.d.). It advises aspiring "creepers" that "there are 'creeportunities' everywhere ... the funnier, creepier or sexier the better" (CreepShots n.d.). As for the women who might be captured in these images, the website has the following advice:

> You can admire the fact that there is something about you sexy enough that somebody wanted to creep you and send it in. If you fall under this category, then maybe leave a comment under the pic and be proud you made it!
>
> (CreepShots n.d.)

With this, the function of women in public existing as objects for the pleasure of men is made explicit – women should be *grateful* that they have "made the cut" and have been subjected to anonymous sexualized commentary.

Third-party use of virtual street-map imagery

Commercial virtual street mapping technologies digitally photograph or otherwise record elements of public space and then recreate them as "explorable worlds" accessible to anyone with an internet connection. The images are collected by cameras attached to the roof of a car that drives along a predetermined route, and then stitched together to create virtual recreations of public space that can be freely and fully explored online (Vincent 2007). The images are taken without regard to what is captured by the camera; virtual street mapping projects simply aim to recreate the world as it is captured, "warts and all," including, then, images of individuals who happen to be present when the car drives by. Such projects have been wildly popular, and justifiably so – they are incredible tools that allow a view of almost anywhere in the world at the click of a button. But there is also a secondary use, in which third parties scour the images for something "interesting" and then share what they find on a separate website. To be clear, all the commercial service providers do is provide the raw imagery – it is the anonymous crowd that seeks out "interesting" photos, cuts them from the hosting website, and then shares them onto other websites for further discussion and commentary. Of course, images can be taken and shared in this manner for journalistic or artistic purposes. Frequently, however, this is not the case. As with r/creepshots, there are countless examples but a few will suffice for illustrative purposes. Users reviewing Street View imagery in England, for instance, found on Google Street View an image that appeared to show a young man vomiting in the street. It was copied and pasted onto another website for discussion, leading one online commentator to argue that it indicated British youth in general were "drunk, jobless, and [with] demonstrably declining IQ," despite no presence of alcohol in the picture (Weisenthal 2008). Viewers of another image that showed three young black men gathered around a car in a relatively poor neighbourhood concluded that they were witnessing a narcotics transaction, despite the complete absence of drugs in the image (Denton 2008). But although this kind of socio-economic or ethnic stereotyping occurs with regularity, it is more often the case that the scrutiny applied is heavily gendered, with a clear focus on sexualized imagery of women.

Of the websites that select (or request users to submit) and then archive "notable" street-level mapping imagery initially taken by Google Street View et al., frequently what is deemed particularly notable is imagery of women that may pique the sexual interest of the audience. For instance, one popular page, entitled "Hot Girls on Google Street View," gathered a collection of 36 photos of women, mostly in bikinis, along with the address of where the photo was taken (TheChive 2014). Two women described as "Half Naked Sunbathing Girls on Google Street View" – again with the inclusion of the residential address where the image was taken – can be found in one prominent archive of street-map images (Google Sightseeing 2007). The poster of that particular image wondered "how these girls feel about being half-naked for all the world to see on Google Street View?" and then asked his fellow website users, "does anyone know who they are?" (Google Sightseeing 2007). Another archive of imagery on a website called "Street View Fun" offers a list of "sexy images";

the photos are *all* of women (Street View Fun n.d.a). "Google Sightseeing" also provides a dedicated archive called "Naked Street View"; the only images that make the cut are those of topless women (Google Sightseeing 2009). These are just a few examples, but all share the same theme: either the operators of the website or (more commonly) its users find images of women somewhere on an existing commercial street mapping service, cut them from that service using a simple screen-capture programme, and then upload the result to the cataloguing website so that they can be viewed and anonymously commented on by other users.

Doxyspotting.com is perhaps the most notable of these services; its sole purpose is to allow its users to share images from street map services of women they believe to be sex workers. Once "found," users upload them along with location information. The "about" portion of the website states the following.

> *Why spot prostitutes on Google Street View?*
>
> Google Maps Street View is the ideal voyeuristic tool. People are voyeuristic in nature – men especially. It is very easy to spot a girl with nice big boobs on Google Street View. Spotting prostitutes may seem a bit more difficult. But if you know what to look for, it actually becomes quite easy. For example you can drop the StreetView Icon pretty much anywhere in Madrid's Zona Industrial de Villaverde and you will be able to spot a hooker on Google Street View. Doxy Spotting is almost a sport. It is the perfect and insuspicious [sic] way to kill some time at work or if you are bored. But be warned – doxy spotting is highly addictive.
>
> (Doxy Spotting n.d.c)

The reference to "sport" encourages competitive behavior amongst users to "hunt down" sex workers and be the first to upload new images to the system. Users are ranked by how many new images they have uploaded, and those with the highest numbers are celebrated in a "top 10" list. Doxyspotting is highly reductive – finding an image of a woman dressed in a particular style or of a particular ethnicity in a particular neighbourhood, and concluding therefore that woman is a sex worker. Users are encouraged to consider women in the "doxyspotting" archive as present or available for purposes of sexual gratification; the website encourages users to use the images as a guide for where they could "plan [their] next holiday," with the implications being clear.

Interestingly, doxyspotting.com appears to have recently adopted the same tactics as have the remaining creepshots-style subreddits in order try and divert scrutiny – its front-page now states that doxyspotting hopes to "raise awareness about working conditions and legality of sex work around the globe," and it includes the logos of the Global Network of Sex Work Projects and the European Network Against Trafficking in Human Beings. However, neither of those organizations link back to doxyspotting.com or mention them on their own websites, and though doxyspotting.com has been active since 2010, the logos were not added to the front-page until April 2015 (Doxy Spotting n.d.a), while the revised mission statement was not added until November of that same year (Doxy Spotting n.d.b). That this statement and these logos exist alongside an FAQ that continues to encourage users to upload images of women with "nice big boobs" casts serious doubt on doxyspotting's claimed mission.

A turn to privacy will not help

The reflexive response to creepshots and certain kinds of third-party use of street-mapping imagery may be to interpret them as invasions of privacy of those captured in the images. However, such framing is problematic for a number of reasons. First, it is very difficult to conceive of a legal action that could capture the type of photography or commentary at issue in this chapter without also rendering actionable a whole host of innocuous behavior or running into (in most jurisdictions) constitutionally protected rights of expression. A hypothetical law that, for instance, banned all public photography without the express consent of every individual captured in the image would curtail the likelihood of someone appearing in an image online without their consent, but would also eliminate the use of Street View by people wishing to see photographs of their childhood neighbourhood or to plan a holiday, severely limit the expressive rights of artists and journalists who rely on street photography for their work, and make it virtually impossible for people to share snapshots of their holiday in Paris on Facebook. Similar challenges would face any attempt to regulate sexualized commentary at any level short of immediate incitement to violence.

The weakness of privacy in the face of unwanted public photography is the reason that participants in the activities discussed herein in fact frequently rely on the language of law *in order to* legitimize their actions. This has the effect of diverting conversation away from any harm suffered by the subject of a photo into a *prima facie* question about whether any particular individual has a legal claim to prevent either the taking of the image or its subsequent circulation. This reliance, essentially, on the conventional legal view that individuals do not have a reasonable expectation of privacy in public, is an attempt by "creepers" to cloak themselves in a veneer of legitimacy. Take, for instance, this anonymous comment left on a website story discussing r/creepshots.

> While there will no doubt be many complaints and moral outrages about this subreddit and how the theme of this subreddit is disgusting, etc … there is absolutely no difference between the types of images that are submitted on r/CreepShots, and between what paparazzi photographers do, around the world, for thousands of newspapers and magazines every single day. Some of the biggest sites on the internet, like TMZ.com rely on candid snaps of celebrities otherwise they wouldn't even exist. Take

> your indignation there first. People constantly have their pictures taken, without their consent, and used in publications without their knowledge. When you are outside and in public space, you do not have a reasonable expectation of privacy.
>
> (Ryan 2012)

The core of this argument is that when photographs are taken in public, the subject has no claim to privacy over that image because anyone present was free to see the same thing. Indeed, this is consistent with tort-based approaches to the protection of privacy in a number of jurisdictions, which generally exclude claims over photographs taken in public places (Prosser 1984; Moreham 2006; Hargreaves 2014). The second argument offered in defence of creepshots by its participants is that the overwhelming majority of images do not actually reveal an identifiable individual, and if there is no identifiable individual then there can be no claim to privacy. This notion also draws from a legal understanding of privacy, in this instance data protection regimes, which typically govern only the collection of "personally identifiable information" (Bygrave 2001). Thus, the aforementioned commercial website that sells access to archives of creepshots allows identifiable individuals to request takedown (removal) of images.

> If however, you wore something sexy, tight or revealing and are shocked, ashamed, belittled or embarrassed that you were creeped and posted on here, then contact us at the email above and include the link and description of the picture. Your request will be reviewed and if found legitimate, the picture will be removed.
>
> (CreepShots n.d.)

(No details are provided as to what exactly qualifies such a request as "legitimate.")

This general approach – an initial contention that, as a matter of principle, privacy cannot be claimed by someone in public space, along with an allowance for the *ex post facto* removal of images of identifiable individuals – also characterizes the approach of the major commercial street map providers, with an added technological twist. Google and Microsoft, the two largest international providers of virtual street maps, both use a combination of automated blurring along with an allowance for individuals to manually request blurring of unblurred imagery or, in certain contexts, request takedown of imagery (Google n.d.; Microsoft n.d.). Though technically impressive, the automated blurring approach is not flawless. Google engineers claim the system used for blurring in Street View now achieves an effective successful blur rate of 89% on faces (Frome et al. 2009). The other service providers have not released data on the accuracy of their blurring efforts, but it is reasonable to assume that they too are striking a similar success rate. When one considers the amount of imagery that these systems are generating, if (at best) one in ten faces goes unblurred before an image is posted, then a significant number of individuals will remain identifiable despite Google or Microsoft's best efforts (Street View Fun n.d.b). At the same time, false positives still pose a problem (Motraghi 2010). Thus, there is a continuing need for the "takedown" option.

However, successful utilization of that option is initially dependent on the individual captured in an image being aware that the image even exists – it is unrealistic to assume that someone will spend time trawling the vast range of websites that archive street mapping imagery in order to determine if there is a photo of them present. Of course, even if they *did* happen to discover an image they wished to have removed, it may have already virally spread beyond the confines of a single website, rendering any takedown request largely ineffective (Linkous 2014). Once an image is placed online, it is effectively impossible to "retrieve" – endless copies can be made and transmitted globally at effectively zero cost. We can see this process play out in both creepshots and third-party use of street mapping photography – images uploaded to one message board or archive might be deleted or removed, but that does not prevent them from reappearing elsewhere. Once an individual has their image propagated via digital means, they must effectively accept that guaranteed permanent removal is a near impossibility – the decentralized nature of the internet actively works against such an outcome (Zittrain 2008).

This means that legal regimes grounded in protecting against "privacy invasions" or "control over information" are likely to be of limited use. But even if they were – that is, even if an individual could successfully obtain a remedy against the individual who uploaded an image without their consent and remove all copies of it from the internet – this would still miss the more significant risk associated with these practices. To understand this risk, we must move away from treating this kind of unwanted photography and commentary as a threat to privacy, and instead consider it more in terms of surveillance.

Surveillance and objectification

Surveillance is a pervasive feature of modern industrial states, where it has become a "general tool used to accomplish any number of institutional goals" (Ericson and Haggerty 2006, 3), ranging from "control, governance, [or] security, [to] profit [or] entertainment" (Los 2006, 73). Modern surveillance, then, "involve[s] the collection and analysis of information about populations in order to govern their activities" (Ericson and Haggerty 2006, 3). For Johnson and Wayland, it is "a set of practices that gather and collect data about individuals or entities, with or without their knowledge or consent, for purposes of an analysis which sorts those individuals or entities on the basis of their behavior or characteristics" (Johnson and Wayland 2010, 23). According to Lyon, the "common denominator [of surveillance practices] is that [they place] people or populations under scrutiny … surveillance tries to make visible the identities or behavior of people" (Lyon 2002, 2), in order to "influenc[e] or manag[e] those whose data have been garnered" (Lyon 2001, 2).

Do creepshots or the kind of third-party use of virtual street-maps discussed in this chapter qualify as such? In other words, do they contain the potential for behavioral management, control, or influence? On the one hand, the women featured in the kinds of imagery discussed herein are often unknown to the individual who captured the photograph and almost certainly unknown to the ultimate viewer. They are usually not identifiable within that image, meaning the collection or dissemination of the image (or secondary re-dissemination in the case of virtual street map images) is not done with the *purpose* of effecting behavioral change in the subject of the image. Instead, the purpose primarily appears to be for the gratification of the viewer. However, I argue that this should not dissuade us from calling these practices "surveillance" because the possibility for behavioral influence still manifests itself in at least two ways. First, surveillance theory would argue that as women become increasingly conscious of the possibility that they may be photographed in public and subsequently subjected to unwanted commentary online, they may seek to avoid such attention by pre-emptively altering their behavior or appearance. Awareness that one may be subject to surveillance may lead to conformity "when perceived or real pressure from others causes us to act differently from how we would act if alone" (Cialdini and Trost 1997, 162). Indeed, ad hoc surveillance increases this pressure, since unpredictable public photography – a car suddenly drives past you with cameras attached to it; a man secretly takes photos of you on your lunch break in a public park – disrupts the ability of individuals to make an informed calculation about their probable exposure in everyday circumstances. When one cannot predict with certainty when a sanction may be applied, who may apply it, or what it may be applied for, the safest approach may be to alter one's behavior across a wide range of activities, at all times. Surveillance may therefore lead individuals to alter an entire range of behaviors, even the most benign. In Cockfield's example, "if an individual thinks that [her] impromptu 'Singing in the Rain Dance' in a downtown centre will somehow be stored and potentially used against her in the future, she may change her behavior and in doing so edit her expression" (Cockfield 2003, 395). Panoptic theories suggest, of course, that this kind of behavioral change may also ultimately become "unconscious," whereby individuals cease to consciously worry about being caught on camera and simply internalize the proper behavior (Foucault 1995; Boyle 1997; Lyon 2001; Gray 2003; Koskela 2003; Yar 2003; Haggerty 2006; Lyon 2006b; Cohen 2008).

However, there is a second (related, but perhaps even more insidious) form of influence at issue with the practices considered in this chapter. It has long been recognized that surveillance is rarely experienced equally, since "different populations are subjected to different levels of scrutiny" (Ericson and Haggerty 2006, 14). For instance, Koskela has argued that though ostensibly neutral in its design, CCTV surveillance reinforces gender disparities, with "women invisible as suspects and as victims, but [present] as targets of voyeuristic use of cameras" (Koskela 2003, 301). Surveillance thus "continue[s] to play an important role in establishing and reinforcing social inequalities" (Haggerty 2006, 29). Ericson and Haggerty have also argued in particular that surveillance can "reproduce sexualized objectifications through the gendered voyeuristic gaze" (Ericson and Haggerty 2006, 15). Lyon too believes that "scopophilia applied to … the viewer society has been translated as a predominantly male gaze that depersonalizes women, turning them into objects to be looked at" (Lyon 2006a, 48). The creepshots phenomena and the repurposing of commercial street map imagery is an example of this kind of gender-based disproportionality in surveillance – a new, digital male gaze. Frederickson and Roberts argue that

> always present in context of sexualized gazing is the *potential* for sexual objectification, [in which] a woman's body, body parts, or sexual functions are separated out from her person … when objectified, women are treated *as bodies* – and in particular, as bodies that exist for the use and pleasure of others.
>
> (Fredrickson and Roberts 1997, 175 [emphasis in original])

Objectification (and its consequent effects) are the second surveillance-related influence we can associate with the practices discussed in this chapter. For viewers of surveillance imagery, Koskela argues, others "are reduced to doll-like bodies lacking personal qualities" (Koskela 2000, 251). These practices are a form sexualized gazing, yes, but is that objectification? Does the gaze discussed in this chapter "separate out" women's body parts or sexual functions from her person? The examples of commentary provided earlier, which focused largely on anatomical features of the subject and subsequent evaluation of sexual desirability, suggests the answer must surely be affirmative. Indeed, when the cropping or blurring serves to largely eliminate the "person" from the image as much as possible the focus is left purely on the physical characteristics that remain: a collection of body parts on display for comment. This kind of objectification primarily "functions to socialize girls and women to, at some level, treat *themselves* as objects to be looked at and evaluated"; over time, argue Frederickson & Roberts, this can lead to "habitual monitoring of the body's outward appearance" (Fredrickson and Roberts 1997, 175). Perceived failure to live up to expected standards of sexual attractiveness can lead to feelings of shame and anxiety, which in turn can manifest in depression, sexual dysfunction, and eating disorders (Fredrickson and Roberts 1997, 181, 186–191; Moradi and Huang 2008, 378; Muehlenkamp and Saris-Baglama 2002, 372). Importantly, this process is experienced not only by those who are actually captured in the images, but rather by all those who are aware of the potential of it happening to them. It is, of course, a form of self-surveillance (Calogero 2012, 575), and so sits comfortably alongside theories of surveillance which postulate behavioral modification as a natural response to unwanted exposure.

While the weight of this new digital gaze falls primarily upon the shoulders of women, the internalization of a particular set of social norms will also impact the surveillers themselves. Those who participate in these communities soon learn that they receive rewards and praise from others for their actions in objectifying women – recall that comments described previously in this chapter that ranged from comparatively innocuous statements discussing the sexual attractiveness of the individual in a photograph, to cruder statements of sexual bravado, to outright threats of sexual violence. Now, it is true that this may be consistent with what Suler has described as the "online disinhibition effect," in which people say and do things online they might not do in "the real world" (Suler 2004, 322–323). However, we should not dismiss the risks as "online only"; the attitudes toward women and the behaviors modelled by the participants are likely to spill over into offline contexts. Those who participate in creepshots activity or those who spend time scouring Street View for images of attractive women may internalize the expectation (or further strengthen an existing expectation) that women substantially exist and perform as the objects of men's sexual interest and desire, rather than as autonomous, self-actualized individuals with interests and desires of their own, unrelated to how they are viewed by men (Galdi, Maass, and Cadinu 2014, 405). Galdi et al. found that "male participants report greater intentions to engage in sexual coercion and are more likely to engage in gender-harassing behavior after viewing [depictions] of objectified women." Similarly, Sterner and Burkley found that men who played video games considered to be "high in perceived sexism" subsequently ranked more highly in "perceptions of rigid gender roles and patronizing attitudes towards women" (Sterner and Burkley 2015, 47–48; see also Dowas and Smith, 2010). Though empirical research regarding the attitudes of those who participate in creepshots-style image sharing online towards gender roles or treatment of women in the offline world has yet to be conducted, this existing literature suggests that a significant connection between the two cannot be discounted.

Solutions: technical and social rather than legal

The solutions for the harms identified in this chapter are unlikely to be found in legal responses aimed at protecting any given individual's privacy for the reasons I have touched upon. Most significantly, as I have indicated, a turn to privacy laws can only address identifiable invasions suffered by identifiable individuals, and even then making out a claim is exceedingly difficult. Moreover, privacy law of course offers no remedy to someone who feels anxiety or shame related to a general culture of objectification. Beyond privacy, other legal regimes are also likely to be of limited use. Take, for instance, copyright law. Google and Microsoft hold the copyright of all the images they publish as part of their respective mapping projects and thus when third-party archives fail to ask for permission to re-publish any of those images, they are violating that copyright. In theory, stricter enforcement of copyright would make it more difficult for third party archives to operate with impunity; however, neither technology giant has shown interest in pursuing copyright claims against the third-party archives for unauthorized use. However, as noted, even if they did the probability of viral image propagation means that the possibility of truly scrubbing the internet of a particular image is virtually impossible. Furthermore, stricter enforcement of copyright would only be salient in the case of commercial street photography of the kind associated with Google and Microsoft; in the context of creepshots, the individual who took the unwanted photograph would in fact be the copyright holder. In general, then, legal responses are likely to be of limited use in solving the problems associated with the forms of image-gathering and sharing discussed herein. Instead, more viable solutions are likely to be technical and social. I will consider each in turn.

On a technical level, service providers could commit further resources to developing techniques that result in lower probability of identification/re-identification of individuals than is currently the case. However, this requires not simply merely a greater level of blurring of faces, since as I have argued, that is a solution that is aimed at the personal privacy of a particular *individual* captured in a photo and the real problem is more systemic. Flores and Belongi have demonstrated an algorithm that could successfully eliminate pedestrians entirely from street imagery without destroying the utility of the street map project (Flores and Belongi 2010). Though this technique does not work well in non-urban areas because it cannot draw on the same kind of surrounding information to make a convincing "fill" of the pedestrian location, Nodari *et al.* have proposed to solve this problem by replacing pedestrians with stock images drawn from an approved data set (Nodari, Vanetti, and Gallo 2012). Assuming technical challenges regarding scale could be surmounted, this would effectively remove the human presence from digital street level maps, thereby eliminating the possibility of street images being repurposed by third parties into mechanisms by which to increase the scrutiny applied to women in public space. However, these technological improvements would, of course, have no beneficial impact on the creepshots phenomena, as they would impact only upon imagery collected by large corporations in the service of their commercial street map programmes.

Ultimately, then, we must turn to a social solution for what is, after all, at its heart a social problem (Rosen 2001). It is notable that the driving factor in the voluntary closure of the original r/creepshots was the revelation by the "Predditors" website of the identities first of the individuals who had uploaded photographs and then of the moderators of the subreddit (perhaps lending further credence to Suler's online disinhibition theory). Though it provoked tremendous anger within reddit as a violation of their community's "no-doxxing" policy, it nonetheless proved highly effective in driving a significant amount of creepshots activity away from reddit to voat (Baker 2012b). Since reddit is one of the most popular websites in the world whereas voat is obscure and has a relatively tiny userbase, this shift meant there were significantly fewer people who simply came across the material by chance. This "social" solution was achieved entirely absent a new legal regime. Indeed, it also suggests that "creepers" are on some level aware that what they are doing is inappropriate, and efforts may usefully be directed at increasing the sense of "social unacceptability" of the practice. The power of internalized social norms cuts both ways – when individuals who are found to be participating in creepshots or similar activity are "called out" by their peers for doing so, it can have a tremendous impact on those around them. It is well-accepted that the significant decrease in "drink-driving" is due as much in part to the "social acceptability" of such behavior as it is to stricter enforcement of laws (Grasmick, Bursik, and Arneklev 1993).

To reiterate, the critical harm associated with these phenomena is not the embarrassment or ridicule that may befall any given individual subject to unwanted photography and subsequent commentary online. Rather, it is the social/systemic message that accompanies the broadly unchallenged circulation of images of women taken in public without their knowledge and/or consent and subsequent sexualized commentary of those images by an anonymous audience. This problem is, therefore, better understood less as a series of discrete invasions of individual privacy, and more as a form of public surveillance – a new digital male gaze. Strengthened social norms regarding the (un) acceptability of surreptitious photography and unwanted image sharing are likely to be the most useful approach to solving the problems discussed herein, and empirical research into this area – particularly in the context of educating children and youth – would be of tremendous benefit. Of course, care must be taken that well-intentioned education campaigns, especially in schools, do not in fact create further problems (Dobson and Ringrose 2016).

Note

1 Title taken in part from Thom Yorke, Jonny Greenwood, Colin Greenwood, Ed O'Brien, Philip Selway, Albert Hammond, Mike Hazelwood. 1992. "Creep", Parlophone Records.

References

Adam, Allison. 2002. "Cyberstalking and Internet Pornography: Gender and the Gaze." *Ethics & Information Technology* 4(2): 133–142.

Baker, Katie. 2012a. "Here's a New, Totally Legal Reddit Hub Devoted to Creep Shots." *Jezebel*, June 27. Accessed April 18, 2016. http://jezebel.com/5921747/heres-a-new-totally-legal-reddit-hub-devoted-to-creep-shots.

Baker, Katie. 2012b. "How to Shut Down Reddit's CreepShots Once and for All: Name Names." *Jezebel*, October 10. Accessed April 19, 2016. http://jezebel.com/5949379/naming-names-is-this-the-solution-to-combat-reddits-creepshots?popular=true.

Boyle, James. 1997. "Foucault in Cyberspace: Surveillance, Sovereignty, and Hardwired Censors." *University of Cincinnati Law Review* 66: 177–206.

Bygrave, Lee. 2001. "The Place of Privacy in Data Protection Law." *University of New South Wales Law Journal* 24: 277–283.

Calogero, Rachel M. 2012. "Objectification Theory, Self-Objectification, and Body Image." In *Encyclopedia of Body Image & Human Appearance*, Vol. 2, edited by Thomas F. Cash, 574–580. San Diego: Academic Press.

Cialdini, Robert B., and Melanie R. Trost. 1997. "Social Influence: Social Norms, Conformity, and Compliance." In *The Handbook of Social Psychology*, 4th ed., Vol. 2, edited by Daniel T. Gilbert, Susan T. Fiske and Gardner Lindzey, 151–192. New York: Oxford University Press.

Cockfield, Arthur J. 2003. "Who Watches the Watcher? A Law and Technology Perspective on Government and Private Sector Surveillance." *Queen's Law Journal* 29: 364–407.

Cohen, Julie E. 2008. "Privacy, Visibility, Transparency, and Exposure." *University of Chicago Law Review* 75: 181–202.

CreepShots. n.d. "About." CreepShots. Accessed April 21, 2016. http://creepshots.com/about/.

Denton, Nick. 2008. "A Drug Deal from Every Angle." *Valleywag*, March 24. Accessed June 12, 2016. http://gawker.com/5004469/a-drug-deal-caught-from-every-angle.

Dobson, Amy Shields, and Jessica Ringrose. 2016. "Sext Education: Pedagogies of Sex, Gender, and Shame in the Schoolyards of Tagged and Exposed." *Sex Education* 16 (1): 8–21.

Dowas, Edward, and Stacy L. Smith. 2010. "Keeping Abreast of Hypersexuality: A Video Game Character Content Analysis." *Sex Roles* 62(11–12): 721–733.

Doxy Spotting. n.d.a. "Project: Doxy Spotting." DoxySpotting.com, April 6, 2015. Internet Archive. Accessed May 16, 2016. https://web.archive.org/web/20150406072644/http://doxyspotting.com/.

Doxy Spotting. n.d.b. "Project: Doxy Spotting." DoxySpotting.com, November 8, 2015. Internet Archive. Accessed May 16, 2016. https://web.archive.org/web/20151108040621/http://doxyspotting.com/.

Doxy Spotting. n.d.c. "Doxy Spotting." DoxySpotting.com. Accessed April 5, 2016. http://doxyspotting.com/about/.

Ericson, Richard V., and Kevin D. Haggerty. 2006. "The New Politics of Surveillance and Visibility." In *The New Politics of Surveillance and Visibility*, edited by Richard V. Ericson and Kevin D. Haggerty, 1–25. Toronto: University of Toronto Press.

Flores, Arturo, and Serge Belongi. 2010. "Removing Pedestrians from Google Street View." In *Proceedings of the 2010 IEEE Computer Society Conference on Computer Vision and Pattern Recognition – Workshops (CVPRW)*, 53–58.

Foucault, Michel. 1995. *Discipline and Punish: The Birth of the Prison*. 2nd Vintage Books ed. New York: Vintage Books.

Fredrickson, Barbara L. and Tomi-Ann Roberts. 1997. "Objectification Theory: Toward Understanding Women's Lived Experiences and Mental Health Risks." *Psychology of Women Quarterly* 21: 173–206.

Frome, Andrea, German Cheung, Ahmad Abdulkader, Marco Zennaro, Bo Wu, Alessandro Bissacco, Hartwig Adam, Hartmut Neven, and Luc Vincent. 2009. "Large-scale Privacy Protection in Google Street View." In *Proceedings of the IEEE*

International Conference on Computer Vision (2009), 2373–2380. doi:10.1109/ICCV.2009.5459413.

Galdi, Silvia, Anne Maass, and Mara Cadinu. 2014. "Objectifying Media: Their Effect on Gender Role Norms and Sexual Harassment of Women." *Psychology of Women Quarterly* 38(3): 398–413.

Google. n.d. "Google Maps Street View: Image Acceptance & Privacy Policies." Google.com. Accessed May 22, 2016. www.google.com/maps/streetview/policies/#privacy-and-blurring.

Google Sightseeing. 2007. "Half Naked Sunbathing Girls on Google Street View." GoogleSightSeeing.com, May 31. Accessed April 5, 2016. http://googlesightseeing.com/2007/05/half-naked-sunbathing-girls-on-google-street-view/.

Google Sightseeing. 2009. "Naked Street View." GoogleSightSeeing.com, March 24. Accessed April 18, 2016. http://googlesightseeing.com/2009/03/naked-people-on-google-street-view/.

Grasmick, Harold G., Robert J. Bursik, Jr., and Bruce J. Arneklev. 1993. "Reduction in Drunk Driving as a Response to Increased Threats of Shame, Embarrassment, and Legal Sanctions." *Criminology* 31(1): 41–67.

Gray, Mitchell. 2003. "Urban Surveillance and Panopticism: Will We Recognize the Facial Recognition Society?" *Surveillance and Society* 1(3): 314–330.

Haggerty, Kevin D. 2006. "Tear Down the Walls: On Demolishing the Panopticon." In *Theorizing Surveillance: the Panopticon and Beyond*, edited by David Lyon, 23–45. Cullompton: Willan Publishing.

Hargreaves, Stuart. 2014. "'Jones-ing' for a Solution: Commercial Street Surveillance and Privacy Torts in Canada." *Laws* 3(3): 388–409.

Herman, John. 2012. "Reddit General Manager Explains Why He Won't Ban Creepy." *Buzzfeed*, October 12. Accessed April 19, 2016. www.buzzfeed.com/jwherrman/reddit-general-manager-explains-why-he-wont-ban-c.

Hern, Alex. 2015. "How Reddit Took on its Own Users – and Won." *The Guardian*, December 30. www.theguardian.com/technology/2015/dec/30/reddit-ellen-pao.

Imgur. n.d.a. "Part II: Creeping through High Schools" (anonymously captured screenshot of r/creepshots thread). Imgur.com. Accessed April 18, 2016. http://imgur.com/a/KkoyV.

Imgur. n.d.b. "/r/creepshots Has Been Removed Due to Doxing of the Main Mod" (anonymously captured screenshot of reddit thread). Imgur.com. Accessed April 19, 2016. http://imgur.com/AL52y.

Johnson, Deborah G., and Kent A. Wayland. 2010. "Surveillance and Transparency as Sociotechnical Systems of Accountability." In *Surveillance and Democracy*, edited by Kevin D. Haggerty and Minas Samatas, 19–33. New York: Routledge.

Koskela, Hille. 2000. "The Gaze Without Eyes: Video-Surveillance and the Changing Nature of Urban Space." *Progress in Human Geography* 4(2): 243–265.

Koskela, Hille. 2003. "'Cam Era' – the Contemporary Urban Panopticon." *Surveillance and Society* 1(3): 292–313.

Linkous, Taylor. 2014. "It's Time for Revenge Porn to Get a Taste of its Own Medicine: An Argument for the Federal Criminalization of Revenge Porn." *Richmond Journal of Law & Technology* 20(4), art. 14: 1–39.

Los, Maria. 2006. "Looking into the Future: Surveillance, Globalization, and the Totalitarian Potential." In *Theorizing Surveillance: the Panopticon and Beyond*, edited by David Lyon, 69–94. Cullompton: Willan Publishing.

Lyon, David. 2001. *Surveillance Society: Monitoring Everyday Life, Issues in Society*. Buckingham: Open University.

Lyon, David. 2002. "Surveillance Studies: Understanding Visibility, Mobility, and the Phenetic Fix." *Surveillance and Society* 1(1): 1–7.

Lyon, David. 2006a. "9/11, Synopticism, and Scopophilia." In *The New Politics of Surveillance and Visibility*, edited by Richard V. Ericson and Kevin D. Haggerty, 35–54. Toronto: University of Toronto Press.

Lyon, David, ed. 2006b. *Theorizing Surveillance: The Panopticon and Beyond*. Cullompton: Willan Publishing.

Microsoft. n.d. "Explore Streetside." Microsoft.com. Accessed May 22, 2016. www.microsoft.com/maps/streetside.aspx.

Moradi, Bonnie, and Yu-Ping Huang. 2008. "Objectification Theory and Psychology of Women: A Decade of Advances and Future Directions." *Psychology of Women Quarterly* 32(4): 377–398.

Moreham, Nicole. 2006. "Privacy in Public Places." *The Cambridge Law Journal* 65(3): 606–635.

Muehlenkamp, Jennifer J., and Renee N. Saris-Baglama. 2002. "Self-Objectification and its Psychological Outcomes for College Women." *Psychology of Women Quarterly* 26(4): 371–379.

Motraghi, Sebastian. 2010. "Picasa, Google Street View, and other over zealous face-recgonition magic." April 1. Accessed April 12, 2016. http://blog.motraghi.us/2010/04/picasa-google-street-view-and-other-overzealous-face-recognition-magic/.

Mulvey, Laura. 1975. "Visual Pleasure and Narrative Cinema." *Screen* 16(3): 6–18.

Nelms, Ben. 2012. "Coweta Teacher Fired for Posting Pics of Students on Creepshots." *The Citizen*, September 28. Accessed April 18, 2016. http://thecitizen.com/articles/09-28-2012/coweta-teacher-fired-posting-pics-students-creepshots.

Nodari, Angelo, Marco Vanetti, and Ignazio Gallo. 2012. "Digital Privacy: Replacing Pedestrians from Google Street View Images." In *Proceedings of the 21st International Conference on Pattern Recognition (ICPR2012)*, 2889–2893.

Predditors. 2012. Predditors.com, October 11. http://predditors.tumblr.com. Internet Archive. https://web.archive.org/web/20121012002055/http://predditors.tumblr.com.

Prosser, William L. 1984. "Privacy [A Legal Analysis]." *Philosophical Dimensions of Privacy: An Anthology*, edited by Ferdinand David Schoeman, 104–155. New York: Cambridge University Press.

Reddit. 2012a. "I am Barack Obama, President of the United States – AMA." Reddit.com, August 29. Accessed April 18, 2016. www.reddit.com/comments/z1c9z/i_am_barack_obama_president_of_the_united_states/.

Reddit. 2012b. "[EFFORT] Reddit's Voyeur Subreddits: Creeping through High Schools." Reddit.com, September 18. Accessed April 18, 2016. www.reddit.com/r/ShitRedditSays/comments/103do2/effort_reddits_voyeur_subreddits_creeping_through/.

Reddit. 2012c. "[META][IMPORTANT] /r/creepshots mod blackmailed into shutting down the sub a.k.a. The Admins Sure Doxxed The Ball On This One." Reddit.com, October 10. Accessed April 19, 2016. www.reddit.com/r/Shi

tRedditSays/comments/119gkw/metaimportant_rcreepshots_mod_blackmailed_into/.
Reddit. 2015. "Admins Have Just Quarantined my 3 year old Subredddit with 66 Thousand Subscribers, /r/CandidFashionPolice." Reddit.com, December 11. Accessed April 19, 2016. www.reddit.com/r/Blackout2015/comments/3wez2y/admins_have_just_quarantined_my_3_year_old/.
Reddit. 2016. "Down to Her Ass." *Reddit.com*, April 12. Accessed April 19, 2016. www.reddit.com/r/VolleyballGirls/comments/4eh9pq/down_to_her_ass/.
Reddit. n.d.a. "r/VolleyballGirls." Reddit.com. Accessed April 19, 2016. www.reddit.com/r/volleyballgirls.
Reddit. n.d.b. "r/YogaPants." Reddit.com. Accessed April 19, 2016. www.reddit.com/r/yogapants.
Rosen, Jeffrey. 2001. *The Unwanted Gaze: The Destruction of Privacy in America*. New York: Random House, Inc.
Ryan, Erin. 2012. "Teacher Posting Creepshots of High School Students to Reddit Rightfully Fired, But Perv Forum Still Exists." *Jezebel* (blog), September 27. Accessed April 18, 2016. http://jezebel.com/5946931/teacher-posting-creepshots-of-high-school-students-to-reddit-rightfully-fired-but-perv-forum-still-exists.
Sarkeesian, Anita. 2016. "Tropes vs Women in Video Games." *Feminist Frequency* (YouTube.com). Accessed April 14, 2016. www.youtube.com/playlist?list=PLn4ob_5_ttEaA_vc8F3fjzE62esf9yP61.
Schleiner, Anne-Marie. 2001. "Does Lara Croft Wear Fake Polygons? Gender and Gender-Role Subversion in Computer Adventure Games." *Leonardo* 34(3): 221–226.
Sterner, S. Paul, and Melissa Burkley. 2015. "SeX-Box: Exposure to Sexist Video Games Predicts Benevolent Sexism." *Psychology of Popular Media Culture* 4(1): 47–55.
Street View Fun. n.d.a. "Sexy." StreetViewFun.com. Accessed April 5, 2016. www.streetviewfun.com/category/sexy/.
Street View Fun. n.d.b. "Unblurred Faces." StreetViewFun.com. Accessed April 12, 2016. www.streetviewfun.com/tag/unblurred-face/.
Suler, John. 2004. "The Online Disinhibition Effect." *CyberPsychology and Behavior* 7(3): 321–326.
TheChive. 2014. "Hot Girls on Street View." TheChive.com, April 2. Accessed April 5, 2016. http://thechive.com/2014/04/02/hot-girls-on-google-street-view-36-photos/.
Vincent, Luc. 2007. "Taking Online Maps Down to Street Level." *Computer* 40(12): 118–120.
Voat. 2015a. "Do I Have to Pretend this is about Fashion Anymore?" Voat.co, December 12. Accessed April 19, 2016. https://voat.co/v/Candid/comments/720955.
Voat. 2015b. "Those Leggings Fit You Fantastic Girl!" Voat.co, December 12. Accessed April 19, 2016. https://voat.co/v/Candid/comments/721308.
Voat. n.d. "Candid." Voat.co. Accessed April 19, 2016. https://voat.co/v/Candid.
Warzel, Charlie. 2015. "Reddit is an Incubator of Hate." *Buzzfeed*, June 26. Accessed April 20, 2016. www.buzzfeed.com/charliewarzel/how-reddit-works-as-an-incubator-of-hate.
Weisenthal, Joe. 2008. "Google Takes Down Pictures of Vomiting British Dude." *Business Insider*, March 20. Accessed July 12, 2017. www.businessinsider.com/google-takes-down-pictures-of-vomiting-british-dude-2009-3.

Yar, Majid. 2003. "Panoptic Power and the Pathologisation of Vision: Critical Reflections on the Foucauldian Thesis." *Surveillance and Society* 1(3): 254–271.
Zittrain, Jonathan. 2008. "Privacy 2.0." *University of Chicago Legal Forum* 2008(1): 65–119.

Chapter 11

Legal standards of location privacy in light of the mosaic theory

Aleš Završnik and Primož Križnar

Introduction

Facebook is using users' phone locations to suggest new friends (Hill 2016). New safety vehicle technologies generate a vast amount of location data in order to operate. The Pokémon Go game is recruiting pedestrians to map the urban world: "It is only a matter of time before it starts selling to businesses the ability to draw in potential customers like that. After that will come location-based advertising, where the phone in your hand will try to sell you things depending on your physical location" (*The Guardian* 2016). The disparate location data – that is, tiles in the mosaic – one leaves behind in the digital age have grown to an unprecedented level (Cimpanu 2016). Seemingly irrelevant data on its own, such as phone logs (Sample 2016) or GPS data from smart phones (Durkee 2010), can reveal sensitive personal data on an aggregate level: where you live, where you work, when you travel, who you meet with, and who you sleep with (Wessler 2015). Mosaic theory builds on the insight that when combined with other data, information becomes much more telling about a person. It suggests that an expectation of privacy exists due to the massive accumulation of information. The theory becomes increasingly important due to the way people live their lives in the digital age and when the possession and use of these devices is not always (meaningfully) voluntary.

According to the European Court of Human Rights (ECtHR) "public information" can fall within the scope of "private life" where it is systematically collected and stored in files held by the authorities.[1] The analysis in this chapter shows that the European and US legal frameworks address the concerns raised by the mosaic theory differently. The US Fourth Amendment jurisprudence has been in great turmoil since the development of location tracking technologies used by law enforcement agencies, while European countries have dealt with these concerns by inventing new rights; for instance, the German Constitutional Court constructed "the right to informational self-determination" (*Recht auf informationelle Selbstbestimmung*). It looks as if European countries do not need the mosaic argument to the same extent as in the United States, although we claim in this chapter that the mosaic theory can become the new legal remedy within the European legal framework. It can further develop the legal standard of reasonable expectation of privacy in public space.

The idea that disparate pieces of information may have no informative value on their own, but can have great value on an aggregate level, is taken up in Europe by other privacy protection instruments as well. For instance, the new General Data Protection Regulation envisages an obligatory Privacy Impact Assessment (PIA) (Article 33) before the employment of mass surveillance of public space. This means that the mosaic argument can curtail disproportionate employment of mass surveillance technologies. Although the mosaic theory is not explicitly mentioned in either the ECtHR jurisprudence or the *acquis communautaire*, its insights are, we claim, implicitly included and should be further elaborated and operationalized.

The chapter will first present the technologies that collect vast amounts of location data. Then it will present the mosaic theory in legal scholarship, before comparing the US Fourth Amendment case law and the European standards of location privacy. It will continue by demonstrating some of the inherent weaknesses of the mosaic argument, and conclude with some recommendations regarding enhanced location privacy protection.

Location-sharing technologies

There are several location tracking technologies that can breach an individual's location privacy. The *Global Positioning System* (GPS) (see Defense Science Board Task Force 2005; Bertagna 2010) consists of a constellation of multiple satellites that can measure location three dimensionally in terms of latitude, longitude, and altitude. By calculating the rate of change in these three dimensions, GPS satellites can also measure the speed and direction of movement. Radio waves are transmitted between satellites and electronic devices with built-in GPS receivers that transmit detailed time and distance (from a specific spot on Earth) data. The process of determining one's position or movement is called trilateration and, for its optimal results, radio waves from at least three Earth orbit satellites must reach the electronic device's GPS receiver. When an electronic device on Earth receives radio waves from at least three separate satellites it performs a trilateration operation, in which the position of the device is determined by drawing imaginary diameters and searching for the intersection of those diameters, which represents the device's location in terms of latitude, longitude, and altitude. If there is a fourth satellite available, the location can be verified again. The result is accurate within approximately one to two meters, whereby the accuracy is even greater if the electronic device determines its position in combination with other systems, such as WiFi.

The *global system for mobile communication* (GSM) (see Article 29 Data Protection Working Party 2011; Bedi 2016) is a digital mobile telephony system operated by national telecommunication service providers (TSPs). They control a specific frequency range, which is divided into smaller areas – cells. Each cell has an identification number (CELL ID), Local Area Cell Code (LAC, which specifies the position of the cell), Mobile Country Code (MCC, from which we can tell the national telecommunication network), or Mobile Network Code (which tells us which TSP controls this cell). Cells are basically towers with antennas providing a digital signal, so that mobile devices can communicate with each other. A mobile device with an inserted SIM card continuously and automatically registers with the closest cell (belonging to the dedicated SIM card TSP) every seven seconds to maintain the strongest signal. This process generates registration data: a record of the closest cell site on which the mobile device is/was connected, the telephone number or Mobile Station International Subscriber Directory Number (MSISDN), the International Mobile Station Equipment Identity (IMEI), and the International Mobile Subscriber Identity (IMSI).[2] When a user starts/ends a call, sends/receives an SMS/MMS, or uses the internet, the device generates data on the initiation, duration, and termination thereof. These communication data between the device and the cell sites include the initial and final time, the duration, and the quantity of communication. The TSP usually stores cell-site location information (CSLI) for diagnostics and billing, but this data can also be used to determine someone's location. TSPs can collect location data on a mobile device and can even send short data packets to the mobile device in order to connect the mobile device to a cell (so-called pinging of the device) and reveal its position.

An *IMSI-catcher* (or *Stingray*) (see Bernstein 2016, Oliver 2013, and Wagner 2014) operates on the basis of the man-in-the-middle attack principle to simulate the GSM cell of the TSP. The IMSI-catcher emits a stronger signal than the TSP cell, thereby causing the mobile device to connect to the IMSI-catcher as the device is programmed to always connect to the strongest signal. Furthermore, the IMSI-catcher can surreptitiously increase the signal strength of mobile devices (and thus interfere with the integrity of the device itself) to determine its position and obtain other information, wherein the cooperation of the TSP is not needed. By such a device, traffic data can be collected too (e.g., IMSI, MSISDN, the date, time, and duration of communications, the amount of data transferred, and also the location of the mobile device). It is a very invasive investigative tool due to its non-selective approach and is often called a *phishing expedition*.

WiFi access points (Article 29 Data Protection Working Party 2011) rely on a unique ID that can be detected by a mobile device. The ID for each WiFi access point is called a MAC (Medium Access Control) address, which is a unique identifier attributed to a network interface. A WiFi access point continuously transmits its own network name and its MAC address, even if no one is using the connection and even in the event the content of the wireless communication is encrypted. Each WiFi access point has its range, so we could say that the radius of a WiFi point represents a cell, which is the spine of internet telecommunications. When a mobile device connects to a WiFi access point, a trail is left in the form of its MAC address, which is "built into" the mobile device. Not only can it be determined that a mobile device was connected to a specific WiFi access point, but the exact location of a mobile device still connected to the WiFi access point can be calculated by means of the Received Signal Strength Indication (RSSI) method.

The *Internet Protocol* (IP) address (Article 29 Data Protection Working Party 2011; Tsou and Lusher 2016) is a unique identification number for each device connected to the internet and some addresses remain the same over time (i.e., static) while others change as a device logs on/off from the network (i.e., dynamic). The IP address can reveal the geographic location of the internet user (IP geolocation) to the internet service provider (ISP) in order to arrange internet service access.

An individual's location or movements can also be tracked with the help of *video surveillance systems* in real time or historically, and this is increasingly done with built-in biometric features. There are many other technologies that can reveal location and, potentially, violate privacy, such as electronic toll collection (ETC), automated number plate recognition (ANPR), systems based on radio-frequency identification (RFID), or automated teller machines (ATMs), but their location gathering capabilities are more transparent at first sight.

The mosaic theory in legal scholarship

The implicit idea behind the mosaic theory is that individual pieces of information might be meaningless on their own but, when taken together, these pieces can reveal patterns of information that convey more than their individual parts. But why has privacy in public spaces not been regarded as raising legitimate concerns? Nissenbaum (1998) convincingly claims that: 1) conceptually, the idea that privacy might somehow be violated in public spaces is often considered paradoxical; 2) one factor contributing to the dismissal of privacy in public is normative in nature (that privacy is regarded as an important value and interest, but it must be balanced against other, competing interests); 3) the empirical status of privacy in public spaces is overlooked because prior to recent advances in information technology, the problem of privacy in public was not experienced in one's everyday life.

The mosaic theory addresses the same concern as the theory of "privacy as contextual integrity." According to Nissenbaum's theory, all of the activities people engage in take place in a "plurality of distinct realms." Behavioral norms vary based on the particular context: in some places it is *appropriate to disclose* certain personal information, but not in other places (e.g. it is appropriate to disclose health-related issues in a hospital, but not in a bank). The context determines what "appropriate" disclosure is. Additionally, in some places, particular personal information can be *freely distributed* while not in others. When both *norms of appropriateness* and *norms of distribution* are respected, there is "contextual integrity" of privacy (Nissenbaum 1997; 2004). The introduction of new practices or technologies in a given context might impact the governing norms of *appropriateness* and *distribution*. These norms could shift, for instance, with vehicle safety communication (VSC) technologies installed in vehicles (Zimmer 2005). For instance, the high level of precision and accuracy of the transmitted identification of vehicles enabled by VSC technology upsets the current norm of appropriate information: what a person can see when looking at traffic is only visual and nonspecific information, while VSC technologies enable continuous around-the-clock digitized data collection. Moreover, VSC technologies might also disrupt the norms regarding the distribution of personal information because VSC vehicles are constantly transmitting information about their identity, location, and status for reception by other vehicles, roadside infrastructure, or anyone else with receiving equipment.

In general, information we leave in public spaces may not be private in itself, but processing it becomes a privacy issue if the governing norms of *appropriateness* and *distribution* related to that piece of information are breached. In other words, generally, the individual does not have a reasonable expectation of privacy in public spaces in discrete circumstances, but if others start to aggregate all data on one's whereabouts, the individual regains such an expectation. The aggregation may qualitatively change the privacy calculation. In this case, the norms of appropriateness (i.e., that only visual and nonspecific information can be collected), are breached and also the mosaic argument can legitimately be raised. In this sense, the mosaic theory complements the "privacy as contextual integrity" theory, as it offers a "unifying" context for how to look at separate pieces of information.

Historically, the mosaic theory first served as an objective opposite to that of the protection of privacy. It was used to explain the operations of the national intelligence agencies (i.e., how they combine disparate sources and data to formulate operative assessments). Freedom of Information Act (FOIA) procedures in cases involving national security in the United States show that the national security agencies have been reluctant to share even *prima facie* irrelevant pieces of information with the public in order to not reveal information on an aggregate level. Such cases included location data and other non-content data. In *United States v. Marchetti*[3], the Fourth Circuit Court used the mosaic theory to support the executive branch, which had decided to stop the publication of a memoir by a former CIA employee by stating: "The significance of one item of information may frequently depend upon knowledge of many other items of information. What may seem trivial to the uninformed, may appear of great moment to one who has a broad view of the scene and may put the questioned item of information in its proper context."[4] But there has been a lot of criticism regarding the secrecy of the government in FOIA procedures, especially after the 9/11 attacks, when the mosaic argument was applied more frequently (Chesterman 2011). This case demonstrates how the mosaic theory can be used very opportunistically, either to support state power, as FOIA cases show, or to empower the individual vis-à-vis the state, as the Fourth Amendment jurisprudence in the next part will demonstrate with the decision in *United Stated v. Maynard*, [5] when the Court made a dramatic shift by claiming that the whole may be more revealing than individual parts.

US jurisprudence on the mosaic theory

The US Supreme Court and the federal Courts of Appeal have made several distinctions in the location privacy jurisprudence about when information gathered by the police culminates in "a search" in terms of the Fourth Amendment and the police should obtain judicial authorization. The core distinctions are: (1) whether the police physically interfered with a suspect's property (trespassed) or not (e.g., in *United States v. Jones*[6], the installation of a GPS tracking device on the suspect's vehicle constituted a trespass and thus a search, implicating the Fourth Amendment); (2) whether the vehicle was parked on private premises when the police installed the GPS tracking device or the vehicle was located in a public place (e.g., *United States v. Marquez* [7] and *United States v. Pineda-Moreno)*[8]; (3) whether a tracked vehicle was owned by the driver or a third person; (4) whether tracking was carried out for a single journey or was continuous for a longer period of time (e.g., *United States v. Knotts*);[9] (5) whether technologically enabled tracking was used as the last resort; and (6) whether the police used a technology that was easily accessible to the public at large (e.g., *Kyllo v. United States*[10]). In tracking the development of the relevant location privacy jurisprudence, the line of development goes through the cases *Katz, Knotts, Kyllo, Maynard*, and *Jones.*

Setting the stage: Katz and Knotts

In *Katz v. United States*, [11] the US Supreme Court decided that the Fourth Amendment protects a person's "reasonable expectation of privacy" and the Fourth Amendment does not solely track property rights and trespass law: "[W]hat [a person] seeks to preserve as private, even in an area accessible to the public, may be constitutionally protected." The fact that the electronic device employed in this case (an electronic listening device attached to the outside of a telephone booth) did not happen to penetrate the wall of the telephone booth had no constitutional significance. The Court criticized the trespass-based rule in several subsequent decisions. Previously, in *Olmstead v. United States*, [12] when the Court established the trespass-based rule, dissenting Justice Brandeis famously wrote that it was immaterial where a physical connection with the telephone wires was made. He argued that the Fourth Amendment should be understood as prohibiting every unjustifiable intrusion by the government upon the privacy of the individual. Similarly, dissenting Justice Murphy in *Goldman v. United States*, [13] claimed:

> [T]he search of one's home or office no longer requires physical entry, for science has brought forth far more effective devices for the invasion of a person's privacy than the direct and obvious methods of oppression which were detested by our forebears and which inspired the Fourth Amendment.

The Supreme Court operationalized the Katz test in *Smith v. Maryland*[14] by establishing a two-pronged test for a search under the Fourth Amendment. Namely, a violation of the Fourth Amendment occurs when the government violates a suspect's *subjective expectation* of privacy that *society* is prepared to recognize as "reasonable." In subsequent cases, the Court substantiated the *subjective expectation* and the social *acceptability* of one's expectations. In one of the leading cases, *Knotts*, the Court decided that there is no reasonable expectation of privacy in public space when traveling in a vehicle on a public road. The Court noted, that such expectation of privacy would not have extended to the visual observation from public places and that nothing from the Fourth Amendment prohibited the police from augmenting their sensory faculties with the help of a beeper. In the circumstances of *Knotts*, this meant there is no reasonable expectation of privacy on a *single* journey. However, this does not cover *all journeys* made over a longer period of time. But the Court decided to not deal with the difficult question of what *all* journeys over a *longer* period of time means and reserved the issue of "dragnet" surveillance for the future.

Sense-augmenting technology

In the subsequent decades, information technology developed to enable continuous, 24/7 tracking, with little cost or effort. In legal terms, the transition from monitoring discrete public movements that are not constitutionally protected, to monitoring public movements that amount to an aggregation of private information that should be protected was yet to be made. The US Supreme Court has tried to draw a line between whether the police can track a suspect without a warrant and still comply with the Fourth Amendment by the introduction of a test of the availability of a particular technology to the public. The difference has slowly grown from a difference of degree to a difference in kind. In *Kyllo*, the Supreme Court did not directly deal with location-tracking technology, but it laid out new criteria of how to review new technologies in terms of a "search" under the Fourth Amendment. The *Kyllo* court dealt with the government's use of an infrared thermal imaging device. These devices measure the temperature of surfaces by measuring the infrared radiation they emit, so if the police wanted to know whether the suspect Kyllo was growing marijuana in his home by using lamps that generate a high level of heat, they did not need to enter the private premises, they only needed to inspect the premises from a public location with a thermal imaging device. The question before the Court was then whether the use of such technology from a public location qualifies as a "search" under the Fourth Amendment. The Court decided that since the technology was "not in general public use," the police should have asked the court for a warrant.

The *Kyllo* test brought new perplexities. Tracking and other information technology follow steep cycles of proliferation and become quickly and easily available to the public at large. The test of the widespread public use of a given technology may thus lead to a different conclusion only a few years later. The *Kyllo* Court built the test on an empirical fact of what counts as a "widespread public use" of a given technology and did not provide clear guidance for the courts. For instance, should the courts in the future resort to a public opinion survey with regard to new technologies or market studies on the proliferation of a particular technology? We argue that the courts should take a more proactive normative role. Perhaps even more important, questions of fact ("what is?") should be distinguished from normative questions ("what ought to be?"). The fact that people widely use spying technology should not lead to the conclusion that law enforcement authorities can freely use such technology. Finally, the *Katz* "expectation-of-privacy" test, as further developed by the *Kyllo* court, is not without its own difficulties as it involves a degree of circularity.[15]

Beeper and GPS data: A difference in degree or in kind?

In several cases, the US Courts have deliberated about whether police use or access to beeper and GPS data counts as a "search" in the meaning of the Fourth Amendment. In the already mentioned *Knotts* case, the US Supreme Court decided that monitoring beeper signals did not invade any legitimate expectation of privacy of a suspect. What about GPS data that can be gathered from afar and without the police physically tailing someone?

The police can harvest GPS data by attaching a GPS device to a vehicle or from third parties, e.g., app producers or internet search engines that have compiled GPS data from their customers during their business practices. Federal and state courts have taken different stances as to whether GPS data is protected by the Fourth Amendment. For instance, in *People v. Weaver*[16], the majority of the Court of Appeals of New York argued:

> Disclosed in [GPS] data … will be trips the indisputably private nature of which takes little imagination to conjure: trips to the psychiatrist, the plastic surgeon, the abortion clinic, the AIDS treatment center, the strip club, the criminal defense attorney, the by-the-hour motel, the union meeting, the mosque, synagogue or church, the gay bar, and on and on.

The New York court envisaged very well what kind of insights can be drawn from aggregated location data and the "substance of life" it can reveal. In

Maynard, the federal Court of Appeals for the District of Columbia followed this logic and made a "dramatic shift" in Fourth Amendment jurisprudence (Gatewood 2014). The court was faced with the following issues: (1) whether the continuous use of a GPS tracking device constituted a "search" under the Fourth Amendment, and (2) if so, whether the search was "unreasonable" in terms of the Katz test. The *Maynard* court rejected *Knotts*'s reasoning by emphasizing the role of privacy: "When it comes to privacy, however, precedent suggests that the whole may be more revealing than the parts." The Maynard court made a crucial step forward by distinguishing two situations: (1) what it means to be *actually* exposed to the public eye and (2) what it means to be *constructively* exposed:

> First, unlike one's movements during a single journey, the whole of one's movements over the course of a month is not *actually* exposed to the public because the likelihood anyone will observe all those movements is effectively nil. Second, the whole of one's movements is not exposed *constructively* even though each individual movement is exposed, because that whole reveals more – sometimes a great deal more – than does the sum of its parts.[17]

There is a difference in kind rather than in degree when the pieces are put together. The Court followed the reasoning that "prolonged surveillance of a person's movement may reveal an intimate picture of his life" and distinguished between "a day in the life" and "a way of life":

> The difference is not one of degree but of kind, for no single journey reveals the habits and patterns that mark the distinction between a day in the life and a way of life, nor the departure from a routine that, like the dog that did not bark in the Sherlock Holmes story, may reveal even more.

However, this view has not prevailed.[18] The Supreme Court reviewed the Maynard decision in 2012 in the retitled action *United States v. Jones*, but the majority did not resolve the case by applying the mosaic theory but instead turned to the trespass doctrine. The *Jones* court decided that attaching a GPS device to Jones's wife's vehicle, which he routinely used, was the crucial point where the police needs to have judicial authorization. The police should have stayed within the limits of the court order to justify infringing Jones's property interest in his wife's vehicle. In his concurrence, Justice Alito strongly repudiated such reasoning by claiming that it is impossible to think of late 18th century situations that are analogous to what took place in the case:

> The Court's reasoning largely disregards what is really important (the use of a GPS for the purpose of long-term tracking) and instead attaches great

significance to something that most would view as relatively minor (attaching to the bottom of a car a small, light object that does not interfere in any way with the car's operation).

Furthermore, the concurring opinions in *Jones* exposed the inadequacy of the persistent *third-party doctrine* in the digital age.

The third-party doctrine and location data

The *third-party doctrine* posits that consumers who knowingly and willingly surrender information to third parties have "no reasonable expectation of privacy" in that information – regardless of the quantity of such information or how revealing it is. In *Smith v. Maryland,* dissenting Justice Marshall argued that, "Privacy is not a discrete commodity, possessed absolutely or not at all. Those who disclose certain facts to a bank or phone company for a limited business purpose need not assume that this information will be released to other persons for other purposes." In fact, the GPS data stored by third parties, such as a smart phone application or a vehicle manufacturer, is another venue for law enforcement to harvest location data and circumvent privacy protection. Several courts have thus fervently rejected the path towards the *third-party doctrine.* The Fourth Circuit Court of Appeals decided the *third-party doctrine* to be not applicable in *United States v. Graham,* [19] wherein the Maryland District Court held that historical cell site location data is not protected by the Fourth Amendment. The Fourth Circuit Court of Appeals wrote: "People cannot be deemed to have volunteered to forfeit expectations of privacy by simply seeking active participation in society through use of their cell phones." This was recognition that sensitive private information can be revealed through users' cell phone records. However, the full court, sitting *en banc*, upheld the Maryland District Court decision in May 2016 and held that the government does not need a warrant to access cell site location data.[20]

In *United States v. Davis,* [21] the records regarding the suspect contained a staggering 11,606 separate location points showing where he was at the start and end of calls – an average of one location point every five-and-a-half minutes, assuming the suspect slept an average of eight hours a night. The Court of Appeals for the Eleventh Circuit did not apply the third-party doctrine and noted that while phone users realize they are giving the phone company the number of the person they are calling, they are generally not aware that they are being tracked. The *Davis* court held that Davis has not voluntarily disclosed his cell site location information to the provider in such a fashion as to lose his reasonable expectation of privacy. Despite finding that the evidence had been obtained in an unconstitutional manner, the Court denied the appellant's motion to exclude the fruits of that electronic search and seizure under the "good faith" exception to the exclusionary rule recognized in *United States v. Leon.*[22] The Supreme Court declined to hear this case on appeal.

In arguing why the third-party doctrine exposes the whole populace to surveillance, Alito (concurring in *Jones*) expressed a crucial insight by claiming that the law of trespass used to be accompanied by practical limitations in the pre-digital age and how these limitations played a significant role in preventing the government from delving too deeply into an individual's personal life. This shows the great importance of the mosaic theory in curtailing the surveillance powers of US law enforcement authorities.

European standards regarding location privacy protection

The mosaic theory helps in overcoming some limits in the Fourth Amendment privacy protection in the US. However, we believe that the mosaic theory can become a new legal remedy in the existing European legal framework due to the established legal standards of Article 8 of the European Convention on Human Rights (ECHR) and the jurisprudence developed by the ECtHR.

The legal foundations of location data protection

According to Article 8 of the ECHR, everyone has the right to respect for his private and family life, his home, and his correspondence and exceptions must be in accordance with the law and necessary in a democratic society for several reasons. An exhaustive definition of the term "private life" does not exist.[23] It is a broad term[24] that is wider than the right to privacy and refers to the sphere within which everyone can freely develop and perfect his personality.[25] That is why ECtHR resolves the question of whether certain information falls within the scope of private life under Article 8 according to the specific circumstances of each case. We claim that location data, which can reveal a great portion of an individual's personality, falls under Article 8. The latter is *prima facie* focused on individuals as the center of protection, and their personal predispositions (such as the integrity of their person, identity, reputation, and life within which they carry out activities and form social links) are protected in the area (which may be of a private or public nature)[26] within which the individual personally develops. The nature of the acquired location data overlaps with the personal predispositions of individuals, because by assembling fragments we can gain insight into their private sphere.

Covert monitoring of movement interferes with the individual's right to freedom of movement too.[27] An individual should be able to move without being monitored and no restrictions should be placed on the exercise of this right other than such as are in accordance with the law and necessary in a democratic society.[28] Location data enables the creation of detailed personal profiles that offer insight into the "darkest corners" of what someone considers private. According to the above, location data is clearly protected by the safeguards of the ECHR and its protocols,[29] but such data may also be gathered by merely referring to the public nature of such information. This specific type of information is subject to legal protection because the individual does not consciously or voluntarily agree to the creation of such a profile, despite the public element of location information.

Methods of obtaining location data and application of the mosaic theory

The ECtHR deems, *inter alia*, the video-surveillance of public places and GPS monitoring[30] to be interferences with the right to respect for one's private life. This type of intervention must be regulated at the legislative level; thus, the mosaic theory cannot be used due to its conceptual nature. If such legislation does not exist, Article 8 of the ECHR[31] is violated, because an interference with the right to privacy is only allowed in accordance with the law (*first scenario*). Despite the various European regulations regarding the use of modern technology for the purposes of law enforcement, protection under Article 8 of the ECHR may be stripped of its import if the regulations do not comply with the ECtHR's standards for legality (*second scenario*). The ECtHR observes that the protection afforded by Article 8 would be unacceptably weakened if the use of modern scientific techniques in the criminal justice system were allowed at any cost and without carefully balancing the potential benefits of the extensive use of such techniques against important private life interests.[32] Contracting States indeed enjoy the competence to determine the conditions under which the system of surveillance is to be operated in the state at issue, but the Court stresses that this does not mean that the Contracting States enjoy unlimited discretion to subject persons within their jurisdiction to secret surveillance – they may not adopt whatever measures they deem appropriate[33] and the law must indicate the scope of any such discretion conferred on the competent authorities or the manner of its application with sufficient clarity to give the individual adequate protection against arbitrary interferences.[34] This is why the phrase "in accordance with the law" directly implies conditions that go beyond the mere existence of a legal basis in domestic law and requires that the legal basis be accessible and foreseeable[35] and therefore formulated with sufficient precision to enable any individual to regulate his conduct, especially due to the fact that the available technology is continuously becoming more sophisticated.[36]

Legal rules are also required to govern the use of the facilities and resources by means of which privacy can be interfered with. These rules should be accessible to the wider public[37] and provide citizens an adequate indication of the circumstances and conditions under which public authorities are empowered to resort to any secret measures.[38] Otherwise, the use of vague and inaccessible legal rules can lead to a different application of the law, arbitrariness of governmental authorities, and inefficient guarantees because the legal basis that led to the gathering of information is not known to an individual or cannot be understood. The legislation should, therefore, contain detailed legal rules to prevent arbitrariness[39] and abuse of the resources of modern technology.[40] If such rules exist but are not precise (e.g., they do not contain the required standard of proof, the duration of the measure, the technological devices to be used, the methods of handling devices, and other necessary restrictions), this would violate Article 8 of the ECHR.

In the *third scenario*, there exists precise and transparent legislation, but law enforcement officers or state authorities bypass it. They use legal methods by means of which crucial data can be gathered, but such methods are not subject to judicial assessment as they are (at least in most European countries) the most common police tasks (e.g. observing and patrolling an area, by means of which location and movement patterns can be ascertained (Gray and Citron 2013), gathering information on someone's location from public announcements on social networks, gathering information about an individual's movements from eyewitnesses and CCTV recordings of public areas, using an IMSI-catcher or thermal-imaging device, gathering information from the web server of a company that voluntarily provides requested data, gathering DNA evidence from a specific place and (in some cases) installing a GPS tracker or gathering data from a vehicle's navigation device without the user's consent for no reason at all. These actions are not subject to *ex-ante* judicial control. By means of the data gathered thereby, an individual's personal profile can be established, from which it may be possible to determine where the individual was or is – as well as his affiliations, beliefs, thoughts, and activities. An observer can therefore create a picture of what individuals seek to maintain as private, or at least hope they will manage to reserve for themselves. This provides an excess of information in the sense that the average observer with his own senses and observations would not be able to create such images and assessments of someone's affairs. Such cases must be resolved by using the legal effect of the mosaic theory by which all such evidence can be excluded as illegal and incompatible with the individual's reasonable expectation of privacy. Such an approach helps in judicial decision-making when a judge assesses whether the use of technology or the conduct of the police have excessively interfered with location privacy in such a manner that the obtained location data reveal intimate aspects of someone's life. In such a case, the judge can exclude the evidence (with regard to which it should cite the protection of privacy) due to the *equilibrium* effect of the mosaic theory or, on the contrary, assume that there was no violation of Article 8 of the ECHR.

The way forward

We assess that, based on the current jurisprudence of the Court, the possibility of applying the mosaic theory already exists. Modern technology as described above can lead to the creation of databases of location data and to such data being stored until it becomes relevant to an on-going investigation. The storage of such data obtained by collecting fragments, and without the knowledge of the person, represents an interference with the private life of the person,[41] regardless of whether such information is ever[42] used for law enforcement purposes.[43] The ECtHR already decided, that systematically stored public information can be subject to protection under Article 8 of the ECHR.[44] Such data is predominantly public and can reveal individual's activities in great detail.[45]

In *Uzun*, the ECtHR established criteria that are relevant in cases in which there arises the question of whether there has been a breach of privacy due to legal actions that created a complete profile of an individual's life. These criteria are the following: the public nature of the gathered information, the time of observation, the methods used in covert actions, systematic collection, the volume of data collection, and the content and nature of the collected data.[46] We can conclude that the criteria established by the ECtHR in the mentioned judgment are substantially the same as those suggested above in relation to the mosaic theory – the public exposure of the location, the grounds required for ordering covert surveillance measures, the authorities competent to permit, carry out, and supervise such measures, the time and scope of the investigation, the methods of observation (with the help of police forces or technology)[47], the content of the information thus obtained, systematic recording, and the scope of the data gathered.[48] Taking these criteria into account, it is possible to determine whether there was a mosaic-equivalent (that is, comprehensive) observation or only fleeting interaction with the individual, allowing an assessment of whether an interference with privacy occurred.

Furthermore, covert surveillance and tracking of an individual's movements may constitute an unlawful interference with their private life[49] if they are not justified (e.g., the protection of national security, the prevention of disorder and crime)[50] and necessary (for the protection of basic democratic institutions and obtaining vital information needed to monitor a specific individual)[51] in a democratic society (Article 8, section 2 of the ECHR). The latter must be assessed by the above-mentioned criteria[52] and, since Article 8, section 2 of the ECHR provides an exception, narrowly interpreted.[53] Furthermore, the Contracting States enjoy a certain margin of appreciation in assessing the existence and extent of such democratic necessity, but this margin is subject to European supervision and the ECtHR has to determine whether the procedures for supervising the ordering and implementation of the restrictive measures are such so as to keep the "interference" within the bounds of what is "necessary in a democratic society."[54]

For verification of the appropriateness and legality of the interventions, the affected person must also have adequate and effective remedies guaranteed that protect him from arbitrary interferences with his privacy.[55] And due to the use of some methods for obtaining location data, legal remedies do not exist until the beginning of a criminal investigation,[56] nor are such means effective, mainly due to the remoteness, the inability to remember one's every step in the past,[57] and the merger of data points collected by one-off interventions, each of which was *prima facie* legitimate. That is why it is necessary to apply the mosaic theory. Only with its application will it be possible to establish effective judicial protection, which lies in the fact that the judge will assess whether there has been an interference with privacy due to the quantity and quality of the data collected, which were individually acquired completely legally, but when merged together form an unlawfully obtained picture of the person being investigated.[58] In the context of such an assessment, not only location data, but also other more sensitive information on an individual's private life obtainable by means of advanced information technology (in the form of communications interception, video recordings, etc.) will be protected. An invasion of privacy in the form of location tracking is considered by the Court to be less invasive and less protected,[59] which is precisely what should not be the case, as demonstrated through the logic of the mosaic. Finally, it is possible to conclude that the mosaic theory approach, through the application of the argument *a minori ad maius*, also protects other data that are existentially related to one's location, disclose one's personal appearance, and can be obtained concurrently by means of modern technology to track one's location.

Criticism of the mosaic theory

The mosaic argument features several weak points of its own. How long is too long for surveillance to count as a "search"? When are we witnessing a qualitative change rather than a quantitative one? These questions are difficult to answer at the abstract level. The courts regularly provide a substantive answer only retroactively, for instance when a court reviews past police activities. Judges will assess a search not by a particular act, but by the aggregate effects of several acts and the conduct of the police as a whole. For instance, while it is clear that the *Knotts* case involved a single trip that was monitored with a beeper, the *Maynard* case involved prolonged GPS monitoring over several weeks, which culminated in a search. The latter means that using the mosaic theory could result in unlimited exclusion of evidence according to the fruits of the poisonous tree doctrine (Kerr 2012).

The strength of the mosaic theory – its flexibility that enables judges to address individual privacy concerns on a case-by-case basis – is also its greatest weakness (Ford 2011). From the perspective of the police, this provides little guidance, as their conduct may be considered illegal if viewed from a wider perspective. This argument of "retroactive unconstitutionality" seems particularly pressing. Individual actions of the police that would otherwise not constitute a "search" may retroactively be labelled as unconstitutional. Because of that, the police must predict if their actions will be considered legal, which is inconsistent with its duty of crime investigation (Gray and Citron 2013).

Pro-mosaic authors have also answered differently the question of how much surveillance is excessive. Gatewood (2014) and Kerr (2012) show how Justice

Alito in *Jones* focused on societal expectations regarding law enforcement practices (i.e., what society is willing to accept as a "reasonable expectation" of privacy), while Justice Sotomayor offered a more normative standard that looked at governmental power, and Judge Ginsburg focused on whether the government learned more than a stranger could have observed.

The question is also what *kind* of search triggers the mosaic theory. Should different methods of location data collection be combined, such as beeper, GPS surveillance, and visual surveillance, or can one method alone trigger the mosaic theory? Kerr (2012) observes that the courts should also consider whether duration and scale questions apply in the same way for every method, as different modes of surveillance have different levels of invasiveness. As a result, he claims, different methods of surveillance might require a different regulation within the mosaic framework.

The mosaic theory also has a negative impact on the allocation of the burden of proof (Ostrander 2011). The accused will have to play an active role in demonstrating that the individual observations created the (whole) image of his private life, so he will have to prove that all criteria for the mosaic theory are met. But this kind of burden can violate the principle of equality, since an excessive burden of proof cannot be considered to offer effective protection of private life, and therefore constitutes a breach of Article 8 of the ECHR.[60]

The lack of more substantiated criteria shows how the mosaic theory sometimes fails to provide clear guidance. Its serves as an "equilibrium-adjustment" tool (Kerr 2011), which means it opens the doors wide for political considerations. The judges adjust the level of protection so that a new technology does not significantly upset the pre-existing balance of power struck at Year Zero (Kerr 2012).

Conclusion

There used to be practical limits to location tracking – neither an average individual nor a well-equipped police task force were capable of observing a person's every movement.[61] Today, it is virtually impossible to fully participate in modern life without leaving a digital trail with location data. These seemingly innocuous data points reveal, when combined, a complete picture of our lives and identities. Tracking individuals is not a social habit and society does not expect (Sloan and Warner 2016)[62] someone's location data to be that systematically gathered and analyzed. However, with the help of tracking technologies and the way of life in digital societies, the sum of location data has increased to an unprecedented level and the value of location data exposed to the public never equals zero. On the contrary, the sum thereof presents a detailed profile of the observed individual.

The mosaic theory claims that pieces of information about one's location may not reveal much on their own, but can reveal "a precise, comprehensive record of a person's public movements that reflects a wealth of detail about her familial, political, professional, religious, and sexual associations" (Sotomayor, concurring in *Jones*). The mosaic theory claims that today location data is widely distributed across traditional borders: from companies to governmental agencies and back, from individuals to companies, and between companies that try to "monetize" our data. In analyzing drones equipped with signal collection devices, Snowden showed the extent that new technologies can harvest location data: if you know

> for example, the different network addresses of every laptop, smartphone, and iPod, you know not just *where* a particular device is in what city, but you know what apartment each device lives in, where it goes at any particular time, and by what route. Once you know the devices, you know their owners. When you start doing this over several cities, you're tracking the movements not just of individuals but of whole populations.
>
> (Snowden 2016)

Therefore, the concept of a "reasonable expectation of privacy" is ill suited to the digital age if it is applied only to discrete, rather than aggregated, forms of observation. Today, individuals do not expect someone without a court order to continuously monitor their public movements and create a detailed picture of their life despite voluntarily revealing their location. Citizens also do not expect law enforcement agents to undertake such actions, as that would run counter to the average life experience (Ford 2011). It is intrinsically contrary to democratic liberal values that location data is known and stored at a detailed level (Tsai, et al. 2010).

The subjective part of the "reasonable expectation of privacy" test should thus be answered positively – one expects location data to be private, whether or not one expects a single location to be private given a one-time exposure. Due to the general expectation that no one is continuously monitoring our public movements without a court order, the objective part of the current "reasonable expectation of privacy" test also still applies. But in order to ensure the existence of both components, it is necessary to use the mosaic theory as a corrective to the obsolete reasonable expectation of privacy concept – as stated by Justice Sotomayor in *Jones*. The mosaic theory is capable of overruling reasonable expectation of privacy exceptions – i.e., the third-party and plain-view doctrines – and providing an effective remedy to safeguard the surplus metadata offered by the "location data assemblage." Without the application of the mosaic theory, law enforcement agencies will use advanced location tracking technology with the justification that they are only gathering publicly available information (Dickman 2011). By such argumentation, the individual's right to free movement will be violated.[63] The mosaic approach therefore safeguards the movements of individuals that take place in the public. Different interests, from law enforcement agencies' pursuit of security to the individual expectation of not being tracked in public, can still be considered proportionally

(Gray and Citron 2013; Kalis 2013). In this way, the mosaic theory further develops the "reasonable expectation of privacy" test in the field of location data and effectively protects an important part of our privacy.

We claim then that the mosaic theory can *reinvigorate* the concept of the "reasonable expectation of privacy," which is widely accepted in the US and Europe. What we learned from Nissenbaum's theory of privacy as "contextual integrity" is that the high level of precision and accuracy of new technologies (e.g., GPS and Vehicle Safety Communication technologies) upsets the current norms of what counts as appropriate data collection (e.g., only visual and nonspecific information that a person can see when viewing traffic) and of appropriate distribution of data. These technologies can also upset the current norms of what can be stored, processed, and distributed freely. The mosaic theory is an alternative way of reinterpreting what counts as "reasonable expectation of privacy" and can be used as an additional argument to the theory of privacy as contextual integrity. The theory of privacy as "contextual integrity" and the mosaic theory both lead to the conclusion that even information exposed to the public can have a claim to remaining private – or better, to regaining the shield of privacy. Assembled pieces of location data can reveal much more at an aggregate level when they form a greater mosaic. To be sure, the mosaic theory has its own limitations. The argument of "retroactive unconstitutionality" seems particularly pressing. But in order to overcome this, pro-mosaic authors convincingly suggest that the police should adopt guidelines and procedures to anticipate the mosaic argument (Gatewood 2014). Law enforcement agencies should take appropriate steps to ensure judicial oversight early enough to avoid evidence having to be excluded at a later stage.

Acknowledgment

Aleš Završnik wishes to express his gratitude to the Eurias fellowship 2017–2018 and Marie-Sklodowska-Curie Actions (COFUND Programme - FP7) programme for awarding a grant that enabled him to research the topic and the host institute Collegium Helveticum in Zürich.

Notes

1 *Haralambie v. Romania*, no. 21737/03, § 95, ECtHR 2010 and *Cemalettin Canli v. Turkey*, no. 22427/04, § 33, ECtHR 2009.
2 In the case of mobile phones, the IMSI is understood as a number that identifies the user by means of an inserted Subscriber Identity Module – SIM card, since the IMSI number is stored on it.
3 466 F.2d 1309, 1318 (4th Cir. 1972).
4 *Id.* § 39. Similarly, *CIA v. Sims*, 471 U.S. 159 (1985).
5 615 F.3d 544, D.C. Cir. (2010).
6 565 U.S. 400 (2012).
7 605 F.3d 604, 610 (8th Cir. 2010).
8 591 F.3d 1212, 1217 (9th Cir. 2010).

9 460 U.S. 276 (1983).
10 533 U.S. 27 (2001).
11 389 U.S. 347 (1967), at 351–352.
12 277 U.S. 438 (1928), at 479.
13 316 U.S. 129 (1942), at 139.
14 442 U.S. 735 (1979), at 740–741.
15 Cf. *Kyllo*, 533 U.S. at 34.
16 12 N.Y. 3d 433, 441–442 (2009).
17 United States v. Maynard, 615 F.3d 544, 558 (D.C. Cir. 2010).
18 See *Pineda-Moreno* and *Marquez*, deciding that the warrantless use of a GPS tracking device on a suspect's automobile did not constitute a search.
19 796 F.3d 332 (4th Cir. 2015).
20 No. 12–4659 (4th Cir. 31 May 2016) (en banc).
21 754 F.3d 1205, 11th Cir. (2014).
22 468 U.S. 897, 1984.
23 *Niemietz v. Germany*, no. 13710/88, § 29, ECtHR 1992.
24 *Costello-Roberts v. The United Kingdom*, no. 13134/87, § 36, ECtHR 1993.
25 *Brüggemann and Scheuten v. Germany*, no. 6959/75, § 55, ECtHR 1997. For example, physical, psychological and moral integrity of individual (*X and Y v. The Netherlands*, no. 21830/93, § 22, ECtHR, 1985), her social identity (*Smirnova v. Russia*, nos. 46133/99 and 48183/99, § 95–97, ECtHR 2003), right to control use of her images (*Reklos and Davourlis v. Greece*, no. 1234/05, § 40, ECtHR 2009), personal reputation (*Polanco Torres and Movilla Polanco v. Spain*, no. 34147/06, § 40, ECtHR 2011) etc.
26 *P.G. and J.H. v. the United Kingdom*, no. 44787/98, § 56–57, ECtHR 2001; *Peck v. The United Kingdom*, no. 44647/98, § 57, ECtHR 2003.
27 Article 2, §1 of the Protocol No. 4 to the ECHR.
28 Set in Article 2, §3 of Protocol No. 4 to the ECHR.
29 See *Uzun v. Germany*, no. 35623/05, § 49–52, ECtHR 2010.
30 See *Peck* at 57 and *Uzun* at 52.
31 See *P.G. and J.H.* at 37–38 and also *Taylor-Sabori v. The United Kingdom*, no. 47114/99, § 16–19, ECtHR 2003.
32 *S. and Marper v. The United Kingdom*, nos. 30562/04 and 30566/04, § 112, ECtHR 2008.
33 *Klass and others v. Germany*, no. 5029/71, § 49, ECtHR 1978.
34 *Dragojević v. Croatia*, no. 68955/11, § 82, ECtHR 2015.
35 This does not mean that an individual should be able to foresee when the authorities are likely to intercept his communications so that he can adapt his conduct accordingly – *id.*, at 81.
36 *Amann v. Switzerland*, no. 27798/95, § 55–56, ECtHR 2000.
37 *Khan v. The United Kingdom*, no. 38030/12, § 27, ECtHR 2000; *Taraneks v. Latvia*, no. 3082/06, § 87, ECtHR 2015.
38 *Weber and Saravia v. Germany*, no. 54934/00, § 93, ECtHR 2006.
39 *Meimanis v. Latvia*, no. 70597/11, § 62, ECtHR 2015.
40 *Savovi v. Bulgaria*, no. 7222/05, § 57, ECtHR 2013.
41 See *Leander v. Sweden*, no. 9248/81, § 48, ECtHR 1987 and *Amann* at 78–80.
42 A 25 year time-limit is in practice tantamount to indefinite retention – *M.K. v. France*, no. 19522/09, § 45, ECtHR 2013, so is 20 years – *Brunet v. France*, no. 21010/10, § 43, ECtHR 2014.
43 *Kopp v. Switzerland*, no. 23224/94, § 53, ECtHR 1998.
44 *M.M. v. The United Kingdom*, no. 24029/07, § 187, ECtHR 2013, at 187, see also *Rotaru* at 43–44 – in these cases the Court states that even pieces of information

about the applicant's life, in particular his studies, his political activities and his criminal record, fall within the scope of "private life" for the purposes of Article 8 of the ECHR.

45 *Malone v. The United Kingdom*, no. 8691/79, ECtHR 1984, see concurring opinion of Judge Pettiti.

46 See *Uzun* at 44, 46, 48–52, 80.

47 *Herbecq and the Association "Ligue des droits de l'homme" v. Belgium*, nos. 32200/96 and 32201/96, ECtHR 1998.

48 Some match criteria generally used in secret surveillance measures – see *Dragojević* at 83.

49 *Shimovolos v. Russia*, no. 30194/09, § 66, ECtHR 2011; similar issues also in *Big Brother Watch and others v. The United Kingdom*, no. 58170/13, ECtHR 2013.

50 *R.E. v. The United Kingdom*, no. 62498/11, § 119, ECtHR 2016.

51 See especially *Szabó and Vissy v. Hungary*, no. 37138/14, § 73, ECtHR 2016.

52 *The Association for European Integration and Human Rights and Ekimdzhiev v. Bulgaria*, no. 62540/00, § 77, ECtHR 2008.

53 *Drakšas v. Lithuania*, no. 36662/04, § 54, ECtHR 2012.

54 *Kennedy v. The United Kingdom*, no. 26839/05, § 154, ECtHR 2010; *Kvasnica v. Slovakia*, no. 72094/01, § 80, ECtHR 2009.

55 See *Weber and Saravia* at 94, *Liberty and others v. The United Kingdom*, no. 58243/00, § 62, ECtHR 2008 and *Roman Zakharov v. Russia*, no. 47143/06, § 236, ECtHR 2015.

56 A refusal of full access to gathered data is necessary where the State legitimately fears that the provision of such information may jeopardise the efficiency of a secret surveillance system – *Segerstedt-Wiberg and Others v. Sweden*, no. 62332/00, § 102, ECtHR 2006.

57 *A simili ad simile M.M.* at 187. ("This is all the truer where the information concerns a person's distant past.")

58 *Pesukic v. Switzerland*, no. 25088/07, § 27–28, ECtHR 2013 – the picture gained through other evidence, which was in itself not sufficient to support an establishment of guilt, but established a strong suspicion, and thus contributed to a full establishment of guilt.

59 See *Uzun* at 52.

60 *Turek v. Slovakia*, no. 57986/00, § 116, ECtHR 2006.

61 The likelihood of that borders on zero – see *Maynard* at 560 and *Knotts* at 278–279.

62 Also *Maynard* at 559.

63 Expectation of privacy is significant even outside of one's residence – *a simili ad simile Perry v. The United Kingdom*, no. 63737/00, § 37, ECtHR 2003.

References

Article 29 Data Protection Working Party. 2011. "Opinion 13/2011 on Geolocation Services on Smart Mobile Devices." 881/11/EN WP 185, May 26. http://ec.europa.eu/justice/policies/privacy/docs/wpdocs/2011/wp185_en.pdf.

Bedi, Monu. 2016. "The Curious Case of Cell Phone Location Data: Fourth Amendment Doctrine Mash-up." *Northwestern University Law Review* 110(2): 61–76.

Bernstein, Henry. 2016. "The Need for Fourth Amendment Protection from Government Use of Cell Site Simulators." *Santa Clara Law Review* 56(1): 177–206.

Bertagna, Patrick. 2010. "How Does a GPS Tracking System Work?" *EE Times*, October 26. www.eetimes.com/document.asp?doc_id=1278363.

Chesterman, Simon. 2011. *One Nation Under Surveillance: A New Social Contract to Defend Freedom Without Sacrificing Liberty*. Oxford: Oxford University Press.

Cimpanu, Catalin. 2016. "If you Clicked Anything Online, Google Probably Knows about it." *Softpedia*, May 19. http://news.softpedia.com/news/if-you-clicked-anything-online-google-probably-knows-about-it-504262.shtml#ixzz49z8uYY00.

Defense Science Board Task Force. 2005. "The Future of Global Positioning System." October 28. www.acq.osd.mil/dsb/reports/2000s/ADA443573.pdf.

Dickman, Bethany. 2011. "Untying Knotts: The Application of Mosaic Theory to GPS Surveillance in United States v. Maynard." *American University Law Review* 60(3): 731–742.

Durkee, Musetta. 2010. "Privacy Expectations in the Use of GPS Tracking Devices: United States v. Maynard." *Berkeley Technology Law Journal* (blog), November 4. http://btlj.org/2010/11/privacy-expectations-in-the-use-of-gps-tracking-devices-united-states-v-maynard/.

Ford, Madelaine. 2011. "Mosaic Theory and the Fourth Amendment: How Jones Can Save Privacy in the Fact of Evolving Technology." *American University Journal of Gender Social Policy and Law* 19(4): 1351–1372.

Gatewood, Jace. 2014. "District of Columbia Jones and the Mosaic Theory – In Search of a Public Right of Privacy: The Equilibrium Effect of the Mosaic Theory." *Nebraska Law Review* 92(3): 504–536.

Gray, David, and Citron, Danielle. 2013. "A Shattered Looking Glass: The Pitfalls and Potential of the Mosaic Theory of Fourth Amendment Privacy." *North Carolina Journal of Law and Technology* 14(2): 381–430.

Kalis, Michael. 2013. "Ill Suited to the Digital Age: Fourth Amendment Exceptions and Cell Site Location Information Surveillance." *Pittsburgh Journal of Technology Law and Policy* 13(2): 1–18.

Kerr, Orin. 2011. "An Equilibrium-adjustment Theory of the Fourth Amendment." *Harvard Law Review* 125(2): 476–543.

Kerr, Orin. 2012. "The Mosaic Theory of the Fourth Amendment." *Michigan Law Review* 111(3): 311–354.

Nissenbaum, Helen. 1997. "Toward an Approach to Privacy in Public: The Challenges of Information Technology." *Ethics and Behavior* 7(3): 207–219.

Nissenbaum, Helen. 1998. "Protecting Privacy in an Information Age: The Problem with Privacy in Public." *Law and Philosophy* 17(5): 559–596.

Nissenbaum, Helen. 2004. "Privacy as Contextual Integrity." *Washington Law Review* 79 (1): 119–157.

Oliver, Nancy. 2013. "Location, Location, Location: Balancing Crime Fighting Needs and Privacy Rights." *University of Baltimore Law Review* 42(3): 485–512.

Ostrander, Benjamin. 2011. "The Mosaic Theory and Fourth Amendment Law." *Notre Dame Law Review* 86(4): 1733–1766.

Sample, Ian. 2016. "Even Basic Phone Logs Can Reveal Deeply Personal Information, Researchers Find." *The Guardian*, May 16. www.theguardian.com/science/2016/may/16/even-basic-phone-logs-can-reveal-deeply-personal-information-researchers-find.

Sloan, Robert, and Richard Warner. 2016. "The Self, the Stasi, the NSA: Privacy, Knowledge, and Complicity and the Surveillance State." *Minnesota Journal of Law Science & Technology* 17(1): 347–408.

Snowden, Edward. 2016. "Inside the Assassination Complex." *The Intercept*, May 3. https://theintercept.com/2016/05/03/edward-snowden-whistleblowing-is-not-just-leaking-its-an-act-of-political-resistance/.

The Guardian. 2016. "The Guardian View on Pokémon Go: Augmented Merchandising." *The Guardian*, July 13. www.theguardian.com/commentisfree/2016/jul/13/the-guardian-view-on-pokemon-go-augmented-merchandising.

Tsai, Janice Y., Patrick Gage Kelley, Lorrie Faith Cranor, and Norman Sadeh. 2010. "Location-Sharing Technologies: Privacy Risks and Controls." *I/S: A Journal of Law and Policy for the Information Society* 6(2): 119–151.

Tsou, Ming-Hsiang, and Daniel Lusher. 2016. "Mapping Web Pages by Internet Protocol (IP) Addresses." *Human Dynamics*, December 18. http://humandynamics.sdsu.edu/document/A2_Mapping%20Web%20Pages.pdf.

Wagner, Stephen. 2014. "Stopping Police in their Tracks: Protecting Cellular Location Information Privacy in the Twenty-first Century." *Duke Law & Technology Review* 12 (1): 201–218.

Wessler, Nathan Freed. 2015. "Taking Warrantless Location Tracking to the Supreme Court." *American Civil Liberties Union*, July 31. www.aclu.org/blog/free-future/taking-warrantless-location-tracking-supreme-court.

Zimmer, Michael. 2005. "Surveillance, Privacy and the Ethics of Vehicle Safety Communication Technologies." *Ethics and Information Technology* 7(4): 201–210.

Chapter 12

Surveillance and privacy in North American public spaces

Bryce Clayton Newell, Silvia de Conca and Kristen Thomasen

Introduction

As discussed in the introduction to this book, there is a longstanding tension between the use of emerging surveillance technologies and personal privacy in public space. Of course, the legal protections for privacy in public space vary by jurisdiction, potentially leading to different experiences in public spaces, and with surveillance in these spaces, across national and local borders. In western academic literature, scholars often draw distinctions between the legal determinations made in different parts of the world about what behavior (or information) is, or can be, protected by privacy law even though it may occur in, or be visible from, publicly accessible places. Despite frequent comparisons between particular jurisdictions or regions of the world (e.g., the United States and western European countries), scholars have not generally paid close attention to comparisons that include all of the three largest jurisdictions in North America: Canada, the United States, and Mexico. Yet these three countries have long shared a political and economic relationship, supported by the frequent cross-border travel of both people and goods. While European countries have long looked to standardizing privacy protection across the continent, privacy standardization has not been the focus of any continental integration in North America. In this chapter, we survey legal protections for privacy in public space across these three countries, focusing largely on the legal regulation of state surveillance.

We hope to develop a better understanding of the similarities, differences, and trends in privacy protection in public space across the continent. Greater comparative understanding in North America can help legislators, courts and academics begin to identify areas where the countries might come together or, at least, where each might learn lessons from one another, in grappling with common challenges raised by state surveillance in public spaces. In this regard, the three countries also share some similarities and exhibit some differences in their legal systems, which makes the comparison more interesting. All three are organized as a federation of states (or provinces and territories). The United States and Canada both primarily share the common-law tradition (with the notable exceptions of Quebec and, in part, Louisiana), while Mexico follows a civil-law tradition. While these countries each rely on different legal provisions, and even different legal systems, we believe that a better understanding of privacy protection (or lack thereof) across the continent can contribute to deeper understanding and development of privacy in public in each of the countries.

For purposes of this chapter, we define (public) surveillance narrowly, following Slobogin (2007), as "government efforts to gather information about people from a distance, usually covertly and without entry into private spaces" (Slobogin 2007, 3). While surveillance can (and should) generally be understood as a much broader subject (see, for example, Lyon 2007, 14; Haggerty and Ericson 2006, 3), this definition allows us to focus primarily on physical surveillance activities conducted by public authorities for purposes related to the investigation or prosecution of criminal or deviant behavior and directed at individuals in publicly accessible places. In this chapter, we rely on a broad understanding of public space that encompasses any place generally held open to access and use by members of the public without significant restrictions, including publicly owned space to which members of the public have a legal right of access. First, we consider the relevant constitutional limits to state surveillance in public space in federal Canadian constitutional law. Second, we present an overview of legal protections for privacy in public in the United States. Specifically, we examine how the Fourth Amendment to the United States Constitution regulates visual/video surveillance and geo-locational tracking by the police. Third, we outline the main provisions concerning the protection of privacy in public in Mexico, highlighting how they interact with cases of surveillance by public actors not traditionally appointed with surveillance duties (civil servants). Finally, in conclusion, we identify and discuss how recent legal developments appear to open opportunities for increased regulation of state surveillance of individuals in public spaces in all three countries.

Canada

Beginning with our northernmost jurisdiction, in this section we explore privacy protections that constrain (or fail to constrain) state surveillance in public spaces in Canada. Several key Supreme Court of Canada decisions set out the parameters for state surveillance in public. Generally, while that which is observable or accessible to the public is ostensibly not subject to privacy protection (and therefore can be surveilled by state agents without limit), at a certain (though yet undefined) point when this state observation is prolonged, or when otherwise anonymous or pseudonymous information about intimate activities is de-anonymized by state actors, some privacy protection will be afforded, as discussed below.

The primary source of legal protection for privacy vis-à-vis state actors in Canada is the constitutional protection arising under section 8 of the Canadian

Charter of Rights and Freedoms.[1] Section 8 affirms that, "Everyone has the right to be secure against unreasonable searches and seizures." The protection of privacy under this provision comes from the interpretation that a "search" takes place when state authorities intrude upon an individual's reasonable expectation of privacy, much like the Fourth Amendment framework in the United States.[2] Section 8 protects privacy proscriptively by requiring state agents to obtain a warrant or other lawful justification for conduct that would invade an individual's privacy. In addition, it protects privacy retroactively by allowing an accused to apply to have evidence obtained from an illegal search excluded from court proceedings, or to seek damages against the state for an illegal invasion of privacy.[3]

To determine whether a state action amounts to a "search," Canadian courts will first determine whether an individual had a subjective expectation of privacy in the subject matter of the alleged search. The court will then assess whether that subjective expectation was reasonable from the perspective of the "reasonable and informed person who is concerned about the long-term consequences of government action for the protection of privacy."[4] In other words, the reasonableness assessment in Canada is normative, rather than descriptive as in the United States. Canadian courts are concerned with the *consequences* of the impugned state action for society, regardless of whether a reasonable person could have anticipated the state surveillance at issue. Accordingly, an individual's awareness that state actors might use technology to collect certain information does not undermine the reasonableness analysis in Canada.[5] Absent a reasonable expectation of privacy, courts will find that a state action does not amount to a search and therefore does not engage constitutional protection. The assessment of whether an individual's expectation of privacy is *reasonable* is therefore typically determinative of the court's analysis.

To assess whether an individual's expectation of privacy in something is reasonable, the courts in Canada look to the "totality of the circumstances" of the alleged search.[6] In the 1996 decision *R. v. Edwards,* and various decisions thereafter, the Supreme Court of Canada began listing some of the circumstances that courts could consider in making this assessment. These, of course, vary depending on the particulars of the alleged search. However, several of the circumstances that courts will consider are particularly relevant to an analysis of privacy in public space, including: the place where the alleged search occurred (i.e. how public was the location of the search); whether the subject matter of the search was in plain view; whether the subject matter had been abandoned; whether the information was already in the hands of a third party and, if so, whether it was subject to any obligation of confidentiality; and whether a surveillance technology exposed any intimate details of the individual's lifestyle or information of a biographical nature.[7] In two key privacy-in-public decisions rendered after *Edwards*, the Supreme Court held that the fact that information was exposed to or accessible in public weighed so strongly against the existence of any reasonable expectation of privacy that these factors were largely determinative.

Specifically, in *R. v. Tessling*, Justice Binnie, writing for a unanimous Court, emphasized that "a person can have no reasonable expectation of privacy in what he or she knowingly exposes to the public, or to a section of the public, or abandons in a public place."[8] The facts in *Tessling* involved the police use of infrared technology to observe heat generated in Mr. Tessling's home. Justice Binnie determined that because the heat was observable on the outside of the home and was therefore perceptible to anyone in public looking at the building (albeit, with the aid of infrared technology), the accused could not reasonably expect any privacy in that heat information – it was publicly exposed. The Court was careful to distinguish these facts from the hypothetical state use of technology to see activities *within* the private home, for which prior judicial authorization would be needed.

Following Justice Binnie's reasoning in *Tessling*, the Supreme Court also held that an individual relinquishes any reasonable expectation of privacy in information contained within property that she has abandoned to the public. In *R. v. Patrick*, the Court had to determine whether an accused had an expectation of privacy in garbage he set out for collection. Given how intimately revealing the information in garbage can be, the Court accepted that individuals generally have an expectation of privacy in household waste.[9] However, Justice Binnie, this time writing for a majority of the Court, held that the reasonableness of this privacy expectation is lost as soon as an accused has done "everything necessary" to designate the garbage for collection – in other words, once the accused has abandoned it to public access.[10] According to the majority, once any member of the public can search the garbage, one can no longer expect privacy in that garbage, including from law enforcement.[11]

Justice Abella dissented on this point, drawing a normative distinction between one's expectation that garbage will be collected and disposed of, and an expectation that it may be searched by state agents.[12] In this dissent, Justice Abella echoes an earlier Supreme Court of Canada decision, *R. v. Wise*, which was decided prior to the "totality of the circumstances" test as set out in *Edwards*, but which was not considered in either *Patrick* or *Tessling*.[13]

Wise involved the use of an electronic monitoring device to track the accused's vehicle for over a month. The majority of the Court held that Mr. Wise had a reasonable, though "markedly diminished," expectation of privacy in his daily movements in his vehicle. One's expectation of privacy on public roads is diminished, according to the majority, because society expects a degree of state surveillance of the roads to ensure public safety. Implicitly though, the majority determined that at some point electronic surveillance goes beyond a reasonable degree of road surveillance, though the decision does not explain when this occurs or how to assess whether this has occurred. The Supreme Court has yet to clarify this analysis.

In separate reasons in *Wise*, Justice LaForest emphasized an important qualitative difference between the risk of being observed by other members of the public, and the risk that an electronic device will track a person's every public movement – a normative distinction that is echoed in Justice Abella's dissenting reasons in *Patrick*, noted above. Justice LaForest explicitly rejects the view set out by the United States Supreme Court in *United States v. Knotts* [14] (discussed on pp.229) that one has no reasonable expectation of privacy once one ventures onto a public street, citing prior Canadian cases which "underlined the potential of uncontrolled electronic surveillance to annihilate privacy" and consequently endanger "individual autonomy and the organization of a free society."[15] He emphasized that the mere fact that one has exposed oneself to other members of the general public cannot on its own justify warrantless state surveillance because "section 8 of the Charter exists to protect privacy and not solitude."[16]

While the reasoning in *Wise* – in particular, Justice LaForest's strong defense of some expectation of privacy in public – was not considered in *Tessling* or *Patrick*, the Supreme Court recently reaffirmed its precedential value in its decision in *R. v. Spencer*.[17] The facts of *Spencer* present a different privacy challenge than *Wise, Patrick*, and *Tessling*. While in the latter cases state agents had a known suspect and sought to gather evidence in relation to his suspected crime, in *Spencer* state agents had collected publicly visible but anonymous evidence of online criminal activity and sought to identify the perpetrator of that crime. The Supreme Court had to decide whether Mr. Spencer had a reasonable expectation of privacy in his personally identifying information held by a third-party internet service provider, which would connect him to his anonymous online activities. Justice Cromwell, writing for a unanimous Court, determined that anonymity is one conceptualization of informational privacy protected by section 8.[18] Accordingly, when assessing the totality of the circumstances of an alleged search, courts can consider whether an accused had a reasonable expectation of privacy by virtue of her anonymity in those circumstances. If she did, a state intrusion will require prior judicial authorization. In practice, law enforcement should be able to obtain such authorization whenever there is evidence of criminal activity. The real impact of *Spencer* on state surveillance is in placing limits on the generalized identification of individuals in public space (i.e. where state agents have no reason to suspect they have committed any crime).

While the facts in *Spencer* dealt with information about Mr. Spencer's online activities, Justice Cromwell adopted Justice LaForest's line of reasoning from *Wise* in asserting that expectations of anonymity can arise in physical public spaces as well. Justice Cromwell explained that

"the mere fact that someone leaves the privacy of their home and enters a public space does not mean that the person abandons all of his or her privacy rights, despite the fact that as a practical matter, such a person may not be able to control who observes him or her in public."[19]

One manner in which those privacy rights may persist in public is through the fact that she is unknown to those who observe her.

Reading the above decisions together it is evident that once someone or something is exposed or abandoned to the public, one cannot reasonably expect privacy from observation, or access to that thing. However, where an individual has a reasonable expectation that her activities are anonymous, state agents will require prior judicial authorization to connect her anonymous activities to her. It is unclear what this requirement will mean in regard to state monitoring of public spaces without identification of individuals (simply observing someone's activity – if she is not a known individual – does not necessarily undermine her anonymity). However, *Spencer* could curb the warrantless use of identifying technologies like facial recognition in public space. Furthermore, at some point, state observation that turns into prolonged monitoring will invade an individual's expectation of privacy and require prior judicial authorization. The Supreme Court has not yet had the opportunity to better clarify the application of *Wise* to long-term public surveillance. However, the seeds have been planted in Justice LaForest's reasons in *Wise*, adapted to an analysis of anonymity in *Spencer*, and in Justice Abella's dissenting reasons in *Patrick*, for an increasing recognition of privacy in public and a normative distinction between exposure to members of the general public, and exposure to state agents. These decisions set out the parameters for the recognition of privacy in public in Canada, and they potentially set the stage for stronger protections for privacy in Canadian public spaces down the line.

United States

In the United States, as in Canada, individuals do not generally have any rights to privacy for behavior or information visible to others in public spaces. The US Supreme Court has repeatedly ruled that individuals, in effect, waive their privacy rights vis-à-vis the police whenever the person (or information about them) enters public space because they have knowingly exposed themselves or their personal information to the public, thus abandoning their legitimate privacy interests in information about them or their activities.[20] However, the US Supreme Court has opened the door to increased privacy protections in public space in a recent case where police tracked a suspect for 28 days using GPS tracking devices. An apparent plurality of the court was open to restricting state surveillance in public when the investigative methods have the capacity to capture "a precise, comprehensive record of a person's public movements" in a way that "evades the ordinary checks that constrain abusive law enforcement practices"[21] and that would allow the state to infer additional sensitive personal information. As such, the "mosaic theory"[22] offers one possible avenue for opening up greater privacy protections in public space in the future. Interestingly, this potential opening from *Jones* developed in a very similar context as that in *Wise*, outlined above. Additionally, the US Supreme Court is currently considering another geolocational privacy case involving the warrantless search and seizure of historical cell phone location records, in *Carpenter v. United States*, and the outcome of that case has the potential to upset existing limitations on the Fourth Amendment's applicability in public space.[23]

In the United States, the Fourth Amendment to the Constitution limits the ability of the state to conduct "unreasonable searches or seizures" without a judicially issued warrant. Surveillance constitutes a "search" (and is thus unlawful absent a warrant) when investigatory action involves a trespass to or interference with a "constitutionally protected area" (e.g., the subject's house, personal property, communications, or body) or when it amounts to a violation of the subject's legitimate (reasonable) expectation of privacy (Whitebread and Slobogin 2015, 98–99). This test ostensibly includes both subjective and objective prongs, as Canada's related test does (although, as noted earlier, the objective prong in the US context is primarily descriptive rather than normative), but federal courts have generally focused only on the objective prong when deciding cases (Kerr 2015). Generally, unaided human observation or video surveillance in a public space will not amount to a search under the Fourth Amendment, especially when the officer (or camera) observes the subject in a publicly accessible place[24] or when the officer observes the subject from a place where the officer has the legal right to be, regardless of where the subject is located.[25] For example, in *United States v. Houston*,[26] the Sixth Circuit held that a person cannot maintain a legitimate expectation of privacy in activities that occur within the area surrounding their home and that are visible to any passersby.

These limitations on the Fourth Amendment's applicability to public-space surveillance are rooted in the premise that a person effectively waives his or her right to (reasonably and objectively) expect privacy in activities that they conduct within the view of others (or in which they expose information to third parties). For example, in *United States v. Knotts*, the Supreme Court held that "a person travelling in an automobile on public thoroughfares has no reasonable expectation of privacy in his movements from one place to another" because "he voluntarily conveyed" that information "to anyone who wanted to look."[27] This reasoning has even prevailed in cases involving visual and video surveillance of the outside of a person's home and yard, whether visible from public streets[28] or from the publicly accessible airspace above the home[29] or other buildings.[30] As the Supreme Court has stated, the Fourth Amendment does not require "law enforcement officers to shield their eyes" as long as they can see the conduct (or object, information, etc.) from where they have the legal right to be, even when "an individual has taken measures to restrict some views of his activities."[31] Similar reasoning has allowed officers to freely rummage through trash bags placed on the curbside outside a home for pickup as, according to the court, placing garbage on the curb amounts to exposing "garbage to the public sufficiently to defeat [a] claim to Fourth Amendment protection,"[32] much like the Supreme Court of Canada's reasoning above.

However, cases do exist that demonstrate how some conduct that occurs in public space can be protected when a person takes reasonable steps to shield their activities from public observation. For example, in *Katz v. United States*, [33] the defendant used a public phone booth to make calls, but the Court held that he maintained a legitimate expectation of privacy that his calls would not be overheard when he closed the door behind him. The *Katz* court maintained the general view that "[w]hat a person knowingly exposes to the public, even in his own home or office, is not a subject of Fourth Amendment protection," but qualified that statement by adding that what a person "seeks to preserve as private, even in an area accessible to the public, may be constitutionally protected."[34] However, this language has not led to much heightened protection in post-*Katz* cases.

The general result of these holdings is that, for Fourth Amendment purposes, most CCTV and other video surveillance of public space is not considered a constitutional "search" at all, precisely because it does not violate expectations of privacy that "society is prepared to recognize as reasonable."[35] Likewise, police use of unmanned aerial vehicles (drones) or body-worn (or hand-held) cameras – either in public spaces or in places where the camera-wearing officer or drone are legally allowed to be – will generally not result in any significantly different legal conclusions under the Fourth Amendment. However, in contrast to the Canadian decision in *Tessling*, the US Supreme Court has held that police use of a thermal imaging device in a public space to measure the heat emanating from the exterior walls of a home violated the Fourth Amendment, "at least where … the technology in question is not in general public use," because it allowed the police to obtain "information regarding the interior of the home that could not otherwise have been obtained without physical 'intrusion into a constitutionally protected area.'"[36]

Some surveillance measures, of course, involve more than merely passive or unaided vision. Technology often augments an officer's natural abilities; for example, video itself can augment memory, zoom lenses magnify vision, thermal imaging or infrared cameras detect radiation not detectable to the human eye, and the use of GPS (and other forms of geo-locational) tracking can augment officers' ability to follow a suspect over long periods of time. A rich body of Supreme Court (and other lower court) case law has addressed some of these questions. For example, cases involving GPS-based tracking or earlier forms of geo-locational tracking have arisen in federal appellate courts since the late 1970s, with the Supreme Court weighing in for the first time in *Knotts*. Many of these early cases, including *Knotts*, involved challenges to the use of "beepers" to track various items, including vehicles, airplanes, and containers or packages of contraband given or sold to suspects. Prior to *Knotts*, courts were divided on the question of whether the Fourth Amendment applied to the use of beepers, although many held that tracking automobiles or airplanes on public roads or airspace, respectively, did not constitute a Fourth Amendment search.

Some of these decisions were predicated on the courts' conclusions that airplanes, automobiles, and contraband were heavily regulated and, as such, they attracted diminished expectations of privacy. Tracking airplane locations was necessary to keep the airspace safe, thus a suspect could not claim a reasonable expectation of privacy in the location of an aircraft. Automobiles moving about on public roads are also exposed to public view and scrutiny, thus diminishing the expectation of privacy a driver or passenger may have in the vehicle's location. Contraband, by its very nature as something that a suspect has no legal right to possess in the first place, cannot attract a legitimate expectation of privacy, thus tracking it cannot violate a Fourth Amendment interest. Some courts also analogized beeper use to visual surveillance, finding no importance (for constitutional purposes) in the augmentation of visual surveillance provided by a transponder. Under these early cases, the strongest case for a Fourth Amendment violation had to do with whether the *installation* of the beeper constituted a search (e.g., it was installed while agents were trespassing or interfering with property in which the suspect had a reasonable expectation of privacy). However, a number of cases did not find Fourth Amendment violations on that basis either.

In *Knotts*, the police had attached a beeper to the inside of a five-gallon container of chloroform with the consent of the company that owned the container. After the suspect purchased the container, the police tracked his vehicle by using both the beeper and by maintaining visual contact until the container was transferred to the second defendant's vehicle. The *Knotts* court held that the use of the beeper did not violate the defendants' Fourth Amendment rights. In the Court's view, the use of the beeper merely augmented the visual surveillance capabilities that the police could generally use, and did not alter their conclusion that surveillance on public roads and in "open fields" did not attract Fourth Amendment protections because the suspects voluntarily exposed their movements in such places to third parties. As stated by the Court:

> Nothing in the Fourth Amendment prohibited the police from augmenting the sensory faculties bestowed upon them at birth with such enhancement as science and technology afforded them in this case.[37]

Although the defendant argued that such a holding would legitimize "twenty-four hour surveillance of any citizen of this country … without judicial knowledge or supervision," the Court stated that "the reality hardly suggests abuse."[38]

A year later, the Court held, in *United States v. Karo*,[39] that DEA agents did not violate a defendants' Fourth Amendment rights when they installed a beeper inside a can of ether prior to its sale to one of the defendants, but that a violation occurred when the agents tracked the can after it entered a private residence. The installation of a tracking device was permissible, the court stated, regardless of whether the agents replaced one of the original cans with one of their own or installed their device into one of the original cans with consent of the owner of the cans at the time the installation took place, because the defendants did not have any legitimate expectation of privacy in the can at the moment of installation (before purchasing it).[40]

Between 1984 and 2012, when the Supreme Court decided *United States v. Jones*, [41] circuit courts around the country continued to decide cases involving different forms of beeper or GPS tracking. Some of these cases extended the reasoning of *Knotts* in the context of GPS tracking, holding that a warrant was not required before inserting a tracking device into an intercepted package and that the installation of tracking devices on vehicles did not violate the Fourth Amendment, because GPS tracking merely utilizes technology to substitute "for an activity, namely following a car on a public street, that is unequivocally not a search within the meaning of the [Fourth Amendment]."[42]

In *United States v. Garcia*, the Seventh Circuit reasoned that "GPS tracking is on the same side of the divide with … surveillance cameras and … satellite imaging, and if what they do is not searching in Fourth Amendment terms, neither is GPS tracking."[43]

Despite the long line of cases described above, one circuit court came to a different conclusion in an influential 2010 decision. In *United States v. Maynard*[44], which later became *Jones* on appeal to the Supreme Court, Judge Ginsburg of the D.C. Circuit Court decided that continuous GPS monitoring of the defendant's vehicle over a 28-day period directly pressed on the question left open by *Knotts* – that is, whether "dragnet-type law enforcement practices" such as "'wholesale' or 'mass' electronic surveillance … requires a warrant."[45] The judge held that prolonged GPS monitoring of a vehicle for 28 days amounted to an unreasonable search, because the GPS monitoring had obtained information that was "not exposed to the public."[46] Judge Ginsburg reasoned that,

> unlike one's movements during a single journey, the whole of one's movements over the course of a month is not *actually* exposed to the public because the likelihood anyone will observe all those movements is effectively nil. Second, the whole of one's movements is not exposed *constructively* even though each individual movement is exposed, because that whole reveals more – sometimes a great deal more – than does the sum of its parts.[47]

This reasoning was informed by the "mosaic theory," which the judge borrowed from case law related to exemptions to disclosure under the federal Freedom of Information Act for national security purposes. According to Judge Ginsburg, prolonged GPS tracking violated the defendant's reasonable expectation of privacy, in part because,

> [p]rolonged surveillance reveals types of information not revealed by short-term surveillance, such as what a person does repeatedly, what he does not do, and what he does ensemble. These types of information can each reveal more about a person than does any individual trip viewed in isolation.[48]

This reasoning was later endorsed at the Supreme Court by both concurring opinions in *Jones*, despite the fact that Justice Scalia's majority opinion avoided the reasonable expectation of privacy test altogether in finding that the installation of the GPS tracking device inside the vehicle's rear bumper violated the defendant's Fourth Amendment rights because it amounted to an unwarranted trespass and interference to the defendant's property (his car). Both Justice Sotomayor and Justice Alito (joined by Justices Ginsburg, Breyer, and Kagan) argued, in separate concurring opinions, that the long-term nature of GPS tracking could also be violation of a defendant's reasonable expectation of privacy. Justice Sotomayor argued, in line with Judge Ginsburg's earlier decision,

> GPS monitoring generates a precise, comprehensive record of a person's public movements that reflects a wealth of detail about her familial, political, professional, religious, and sexual associations. The Government can store such records and efficiently mine them for information years into the future. And because GPS monitoring is cheap in comparison to conventional surveillance techniques and, by design, proceeds surreptitiously, it evades the ordinary checks that constrain abusive law enforcement practices: "limited police resources and community hostility."[49]

There are some striking similarities between *Jones* and *Wise*, particularly as both cases rely on relatively long-term location tracking to open up possibilities for stronger protections for personal privacy in public spaces in future decisions. In *Wise*, the Supreme Court of Canada spent more time analyzing whether such surveillance impinges a person's reasonable expectation of privacy (finding that it does), but the concurrences in *Jones* provide more substantive tools by which to assess when reasonable expectations of privacy are engaged. Thus, these cases have set the stage for future decisions, such as *Carpenter v. United States*, that may ultimately determine whether continuous or pervasive surveillance of public spaces (or people moving about in public spaces) over periods of time will ultimately generate greater constitutional protections for personal privacy in public space.

Mexico

While the Canadian and US supreme courts have developed privacy-in-public jurisprudence by interpreting and applying constitutional protection in a common law system, Mexico's entire legal system belongs to the civil law tradition, with an important role played by the Federal Constitution[50] and the federal codes. As such, Mexico provides for an interesting case for comparison within the North American context. Mexico has a federal structure, like the United States and Canada, that consists of 31 states and the Federal District of Mexico City. Existing laws throughout Mexico primarily focus on regulating fixed video cameras installed on or near streets and public squares (in addition to some data protection regulations discussed on p.000). Since 2012, in fact, public camera surveillance has seen a constant and exponential growth, supported by public opinion and governmental funding programs, such as the Bicentenario Plan, SUBSEUM 2015 and FORTASEG 2017, and is carried out in all the Mexican states, regardless of the existence of regulation (Arteaga Botello 2015). Unlike their neighbors to the North, Mexican Courts do recognize a core of individual privacy protection – including in public. In this regard, Mexico has quickly developed a set of judicial decisions that draw from the tradition of Spain and other European countries. At the same time, technological development within the country has forced the government to face new and important challenges that highlight the possibility for new legislative interventions. After briefly explaining the relevant legislative framework, we outline how the Supreme Court of the Nation has balanced conflicting interests regarding privacy and surveillance in public spaces and discuss how Mexican law could provide greater protections for privacy in public space.

On one hand, protecting the public order has a constitutional basis under Mexican law – indeed, Article 21 of the Mexican Charter affirms that public security is a function of the federal, state and municipal authorities. On the other hand, the right to privacy is not explicitly regulated by the Mexican Constitution and, as such, privacy is not explicitly enumerated as a constitutional right. However, in 2008, following a long debate, the Second Section of the Supreme Court of Justice of the Nation (*Suprema Corte de Justicia de la Nación*) connected the right to privacy to Article 16 of the Mexican Constitution. Article 16, which establishes the right of integrity of the body, family, domicile, communication, and property of all individuals, has been interpreted extensively by the Supreme Court, connecting the inviolability of the domicile to the protection of the intimate and private life of individuals and their families.[51] Subsequently, the Supreme Court affirmed that the protection of privacy exists at two different levels: one where privacy is protected from interferences by other persons and another where the privacy protection is exercised against the State itself. For this reason, the government cannot disregard privacy rights except in exceptional cases when the state can provide objective and reasonable justifications that are not abusive or arbitrary.[52] In the same year, the Collegiate Circuit Court also affirmed – in an attempt to balance conflicting interests – that being photographed in the street as part of an ongoing police investigation did not constitute an interference with Article 16 of the Constitution due to the prevailing interest in public security. As explained by the Court, "in these cases the interest connected to the investigation of an illicit action gains more relevance."[53]

The general principle underlying such decisions is that while individuals have the right to privacy vis-à-vis state authorities, especially in the context of the family, domicile, property, or with regard to their communications (those elements expressly included within Article 16), such a right must be weighed and evaluated against competing interests in public order and security to understand whether a subject has a reasonable expectation (*expectativa legítima*) to privacy.[54] For this reason, surveillance within a public space by the authorities can be legitimate if carried out, for example, in connection with a criminal investigation, or if carried out for security reasons, without abusing authority or power. When surveillance might become abusive, however, must be decided on a case-by-case basis. Due to the constitutional nature of the right to privacy, even temporary or short-term surveillance can constitute a breach of the right when its implementation lacks a legitimate basis under the law. Prolonged surveillance activities, such as the tracking of an individual based on the location of their mobile phone, are also permissible when they are based on a legitimate basis, such as those discussed on p.000 (Gutiérrez Ortiz Mena 2014).

Data protection law now also plays an important role in protecting privacy in Mexico. In January 2017, the federal Law for the Protection of Personal Data in the Possession of "Subjectos Obligados"[55] entered into force. The law protects personal data collected and processed by governmental, federal, and local authorities – like existing provisions related to personal data processing by private parties. Article 6 of the law, however, also expressly limits these protections in the contexts of national security and public order. Therefore, the law does not apply to personal data collected through surveillance in public when the surveillance is used as a tool to maintain either of these state interests.

Within the federal framework, the governments of four states, the federal district, and two municipalities have also issued laws regulating the employment of technological means to pursue public security. As of the first quarter of 2016, surveillance in public was regulated in Mexico City, the states of Mexico, Aguascaliente, Colima, Durango, the Municipality of Guadalajara, and the Municipality of Sayula (both in Jalisco) (Ibarra Sánchez 2012; Arteaga Botello 2016). Regulations in these jurisdictions, except for the Federal District, establish that the areas to be surveilled must present situations of reasonable risk and/or concrete danger, and the criteria of proportionality – meaning both that the use of video surveillance must be appropriate and that it must only interfere with the right to intimacy, privacy, and honor to a minimal degree (Arteaga Botello 2016; Téllez Valdés 2012). This appears to afford much stronger protection of privacy in public space than offered by the Canadian and US constitutions.

The right to intimacy, defined as the personal space of every individual necessary to develop their personality, can also be protected from interference in public under the condition of anonymity (Suprema Corte de Justicia de la

Nación 2016) – a similar protection to what has also recently developed in Canada. However, intimacy and privacy rights are also subject to important limitations based on the state's interests in public order and security. For example, being photographed or filmed in the streets does not constitute a violation of the right to intimacy and privacy when such actions are part of an investigation of a crime, even if the subjects filmed or photographed are not the suspects, and even more so if the material is only collected for use in the context of a criminal trial.[56]

The Mexican Supreme Court has defined honor as the reputation of individuals, strictly connected to ethical, social, and moral values, and divided it into a subjective element (personal perception) and an objective one (perception of third parties). In this sense, the Court recognized that even in the absence of any damage to the objective element of honor, damage could still occur to the subjective aspects when an individual feels that their dignity has been damaged.[57] In all the cases mentioned above, the evaluation is to be carried out on a case-by-case basis.

Finally, privacy and personal data protection also play a role in how the audio-visual information collected through surveillance measures is managed and disclosed to the public. Most of the local laws consider: 1) access to the information under the Law on Transparency and Access to Governmental Public Information; 2) the erasure or redaction of faces, plate numbers, or other personal information from the recordings, but only when such information is not necessary for the investigation or prosecution of crime and the subjects are third parties not involved in the recorded crime; and 3) the automatic destruction of information after a certain lapse of time (between 30 and 180 days) if it is not deemed useful or necessary to an investigation or judicial proceeding.

However, since the institutions collecting the material sometimes have the duty to transfer the material collected to other public organisms, the deletion of such records might occur within the collecting public body (whether upon request of the parties involved, or following the expiration date of the material), but other institutions might continue to retain such material, significantly diminishing the protection (Téllez Valdés 2012; Ibarra Sánchez 2012).

Despite the numerous layers of applicable federal, state, and local law, the Mexican legal framework for regulating surveillance in public space exhibits a number of deficiencies. For example, while the local laws explicitly consider fixed cameras, no reference is made to movable cameras. In this way, the use of body cameras and car-mounted cameras by police officers, recently announced by the public authorities of the Federal District, might not necessarily be covered by the existing legislation, and could take place in a legislative lacuna. These and additional gaps are exemplified quite clearly by a peculiar case that occurred in Mexico City in 2015 and 2016, when a civil servant repeatedly filmed citizens while they were in the act of violating the law and live-streamed these recordings through Periscope.

In February 2015, Arne Aus den Ruthen Haag, a former politician and newly appointed City Manager of the Delegaciòn Manuel Hidalgo, one of the 16 administrative districts of Mexico City, announced the creation of the *Operativo Vecino Cochino* (Dirty Neighbor Task Force), a team of civil servants, sometimes accompanied by public forces, in charge of patrolling the streets of the district to prevent the occurrence of minor administrative violations, with a special attention to littering, traffic, and parking violations. On his first day patrolling, Aus den Ruthen Haag filmed a woman in the act of littering and live streamed the entire scene online using the Periscope app on his mobile phone (using his personal account). The video immediately went viral, and Aus den Ruthen Haag subsequently uploaded the recording to his personal YouTube channel, giving life to the trending hashtag *#LadyBasura* (in English: *#LadyGarbage*). In the following months, Aus den Ruthen Haag and his Task Force kept patrolling the streets of the Manuel Hidalgo, from time to time live streaming on Periscope their most heated and tense encounters (which were shared and followed by tens of thousands of Internet users). According to Aus den Ruthen Haag, he streamed about 2% of his interventions to ensure the correct and transparent behavior of civil servants (himself and his task force), rather than to shame the subjects involved. In mid-February 2016, the Commission for Human Rights of Mexico City started investigating three complaints from citizens who had been filmed and streamed on Periscope by the city manager. The Commission affirmed that since personal images represent personal data, the authorities had to respect the obligations established by the law, stressing its concern for the possible damages caused to the citizens involved due to the violent reactions and the threats recorded online, and to the lack of proportionality between the online shaming and the (lack of) seriousness of the infractions. The Commission also requested a technical opinion from the Institute for the Access to Public Information and the Protection of Personal Data of Mexico City to establish whether the use of Periscope by civil servants could constitute a violation of the law (see CDHDF 2016).

The conduct of the city manager and his team was supported by the head of the Delegaciòn Manuel Hidalgo, who live-streamed a declaration affirming that their conduct was protected by the right to expression, that the use of online platforms was a way to ensure transparency, and that the conduct also helped prevent the violation of the fundamental right to safety and security of other citizens. Acknowledging the lack of relevant legislation, the head of the Delegaciòn affirmed that her department was going to keep using Periscope and YouTube. However, she did not include any guidelines for protecting the privacy of those subjected to the surveillance. Additionally, the city manager stated in an interview released to the activist website Animal Politico in 2016 that those acting in public and, in particular, those acting against the law, had renounced to their right to privacy. In mid-April 2016, the activities of Aus den Ruthen Haag were temporarily interrupted by decree by the President of the Legislative Assembly of Mexico City as contrary to the provisions regulating electoral campaigns. Finally, in June 2016, the city manager withdrew from his office following an administrative procedure started against him for events unrelated to his use of Periscope.

Notwithstanding the lack of a proper conclusion to the events, this case represents a turn in the relationship between privacy and public surveillance in Mexico. It has opened the debate around the use of new technologies for public security, involving not only scholars and experts, but also public authorities at various levels of government and concerned members of the public. Importantly, the case has highlighted how these new tools might empower sectors of the public administration that were traditionally not capable of carrying out public security tasks and that may not be adequately trained to do so in a way that is respectful of all the interests involved. It also suggests that Mexico's data protection law might provide a legal mechanism for regulating some forms of state surveillance and, potentially, providing greater protections for privacy in public spaces in Mexico.

Conclusion

Although all three jurisdictions regulate surveillance and privacy differently, we do see some similar trends emerging. In particular, all three countries are increasingly grappling with similar questions about when (or whether) to legally protect some degree of privacy in public space. Generally, there is little (if any) privacy-based protection for activities or personal information that are exposed to public view in public, or publicly accessible, space, at least as it relates to government surveillance for criminal investigation purposes. All three jurisdictions rely, at least to some extent, on a reasonable (legitimate) expectation of privacy test. The reasonable expectation of privacy test in Canadian Section 8 jurisprudence encompasses subjective and objective inquiries conducted through a normative lens. In the United States, the reasonable expectation of privacy test has become primarily a descriptive, objective test, despite some cases, like *Katz*, highlighting the more subjective prong of the test when analyzing the active effort by a subject to protect their privacy and anonymity in public. Mexico's application of the reasonable expectation of privacy test under Article 16 of the Mexican Constitution is also essentially an objective test used to balance privacy and security interests. Furthermore, while Canadian and US courts often inquire into the intrusiveness of surveillance (for example, looking at the duration or scope of surveillance, the nature of the space where the surveillance occurred, or the type of personal information collected), the civil law tradition in Mexico treats these elements as irrelevant. In Mexico, locations do not assume any special relevance, as all Constitutional rights, privacy included, are always linked exclusively to the individual. Importantly, the data protection analysis under Mexican law does not encompass a reasonable expectation of privacy element. Rather, it focuses directly on regulating the collection, processing, and management of personal data by public authorities. In this sense, Mexican law (with its data protection approach), resembles the approaches taken throughout continental Europe more closely than those taken by the United States and Canada,[58] imbuing data protection considerations into the legal analysis of surveillance by state authorities in public spaces.

In both Canada and the United States, courts have traditionally held that individuals essentially waive (or abandon) their legitimate expectations of privacy vis-à-vis public authorities (and, for most purposes, private actors as well) when they enter public spaces. In Mexico, while courts have held that individuals can maintain some legitimate expectations of privacy in public space, the right is significantly limited in practice by other, prevailing interests (such as the interest to public order and security), provided that the authorities do not abuse their powers. More recently, in Canada, *Spencer* recognized some right to anonymity in Canadian public spaces, standing for the proposition that merely entering a public space should not negate any and all reasonable expectations of privacy. And the respective Supreme Court decisions in *Wise* and *Jones* open the door for increased recognition of legitimate expectations of privacy in public space in the context of long-term location tracking in both the Canadian and US contexts. In Mexico, data protection law may provide one way to obtain some protection from state surveillance (or at least, certain disproportionate further uses and processing of personal data captured in public spaces), especially when, as seen above, constitutional protections may also be overridden by objective and reasonable justifications, but it is not clear yet how strong this protection will be in the future.

Our analysis in this chapter does not cover all relevant state, provincial, or local-level legislation and case law, so there will be variation in the levels of protection offered in some parts of these large countries. Generally, however, the limited nature of the right to privacy in public space in all three countries appears to be driven, as least in part, by tensions between the publicly visible and accessible nature of "public spaces" themselves and lingering notions of privacy as something equivalent to secrecy[59] (a position criticized by Justice Sotomayor in *Jones*), as well as significant tensions between privacy rights, on one hand, and societal interests in freedom of expression, law enforcement/criminal justice, and public order and security, on the other. However, in the end, we find that, in each of these three countries, recent legislative and judicial decisions have opened the way for increased protections for privacy in public space. The next question will be to what degree these countries look to each other's experiences to drive future developments in the law.

Notes

1 Part I of the Constitution Act, being Schedule B to the Canada Act 1982, ch. 11 (U.K.). See also Privacy Act, R.S.C., ch. P-21 (1985).
2 *Hunter v. Southam*, [1984] 2 S.C.R. 145 (Can.), pg. 159.

3 *Charter*, ss. 24(1) and (2); *R. v. Collins*, [1987] 1 S.C.R. 265; *Ward v. City of Vancouver*, [2010] 2 S.C.R. 28.
4 *R. v. Patrick*, [2009] 1 S.C.R. 579, para. 14 (Can.). See also *R. v. Spencer*, [2014] 2 S.C.R. 212, para. 18 (Can.).
5 *R. v. Tessling*, [2004] 3 S.C.R. 432, para. 42 (Can.); *Spencer*, 2 S.C.R. 212 at para. 18.
6 *R. v. Edwards*, [1996] 1 S.C.R. 128, para. 45 (Can.).
7 *Tessling*, 3 S.C.R. 432, para. 32.
8 *Tessling*, 3 S.C.R. 432 at para. 40.
9 *Patrick*, 1 S.C.R. 579 at paras. 2, 32.
10 *Id.* at paras. 25, 55, 62.
11 *Id.* at para. 40.
12 *Id.* at para. 89.
13 *R. v. Wise*, [1992] 1 S.C.R. 527 (Can.). *Wise* is only mentioned in *Tessling*, 3 S.C.R. 432, at para. 22.
14 *United States v. Knotts*, 460 U.S. 276 (1983).
15 *Wise*, 1 S.C.R. at 563, citing R. v. Duarte, [1990] 1 S.C.R. 30 and *R. v. Wong*, [1990] 3 S.C.R. 36.
16 *Wise*, 1 S.C.R. at 564.
17 *Spencer*, 2 S.C.R. 212 at paras. 43–44.
18 *Id.* at para. 38.
19 *Id.* at para. 44.
20 *Katz v. United States*, 389 U.S. 347, 351 (1967); see also *California v. Greenwood*, 486 U.S. 35, 40 (1988).
21 *United States v. Jones*, 565 U.S. 400, 415–16 (2012) (Sotomayor, J., concurring).
22 *United States v. Maynard*, 615 F.3d 544, 562 (D.C. Cir. 2010), citing *CIA v. Sims*, 471 U.S. 159, 178 (1985).
23 *United States v. Carpenter*, 819 F.3d 880 (6th Cir. 2016), cert. granted by *Carpenter v. United States*, 137 S.Ct. 2211 (Mem) (2017).
24 *Knotts*, 460 U.S. at 281–82.
25 *California v. Ciraolo*, 476 U.S. 207, 213 (1986); see also *United States v. Jackson*, 213 F.3d 1269, 1280 (10th Cir. 2000).
26 *United States v. Houston*, 813 F.3d 282 (6th Cir. 2016).
27 *Knotts*, 460 U.S. at 281–82.
28 *Houston*, 813 F.3d 282.
29 *Ciraolo*, 476 U.S. 207.
30 *Dow Chemical Co. v. United States*, 476 U.S. 227 (1986).
31 *Ciraolo*, 476 U.S. at 213.
32 *Greenwood*, 486 U.S. 35.
33 *Katz*, 389 U.S. 347.
34 *Katz*, 389 U.S. at 351.
35 *Kyllo v. United States*, 533 U.S. 27, 34 (2001).
36 *Id.*
37 *Knotts*, 460 U.S. at 282.
38 *Id.* at 284.
39 *United States v. Karo*, 468 U.S. 705 (1984).
40 *Karo*, 468 U.S. at 712.
41 *United States v. Jones*, 565 U.S. 400 (2012).
42 *United States v. Cuevas-Perez*, 640 F.3d 272, 273–74 (2011).
43 *United States v. Garcia*, 474 F.3d 994, 997 (7th Cir. 2007).
44 *Maynard*, 615 F.3d 544.
45 *Maynard*, 615 F.3d at 556–58.

46 *Maynard*, 615 F.3d at 558.
47 *Id.*
48 *Maynard*, 615 F.3d at 562.
49 *Jones*, 565 U.S. at 415–16.
50 Constitución Política de los Estados Unidos Mexicanos [Const.], as amended, Diario Oficial de la Federación [D.O.], 5 de Febrero de 1917 (Mex.).
51 Derecho a la Privacidad o Intimidad. Está Protegido Por el Artículo 16, Primer Párrafo, de la Constitución Política De Los Estados Unidos Mexicanos, Suprema Corte de Justicia [S.C.J.N.] [Supreme Court], Semanario Judicial de la Federación y su Gaceta, Novena Época, tomo XXVII, Mayo de 2008, Tesis 2a. LXIII/2008, Página 229 (Mex.).
52 Libertad de Expresión y Derecho a la Información. Forma en Que la "Moral" o "Las Buenas Costumbres," Pueden Constituir Restricciones Legítimas a Dichos Derechos Fundamentales, S.C.J.N., Gaceta del Semanario Judicial de la Federación, Décima Época, libro 3, Fedrero de 2014, Tesis 1a. XLIX/2014, Página 672 (Mex.).
53 Derecho a la Intimidad o a la Vida Privada. No se Viola Por el Hecho de Que el Imputado o Algún Miembro de su Familia Sea Fotografiado en la Vía Pública Como Parte de las Labores de Vigilancia e Investigación Para la Persecución del Delito Correspondiente, Primer Tribunal Colegiado del Segundo Circuito Con Residencia en Ciudad Nezahualcóyotl, Estado de México, Gaceta del Semanario Judicial de la Federación, Décima Época, Tomo II, Libro 20, July 2015, Tribunales Colegiados de Circuito, p. 1719, Tesis: II.1o.29 P (10a.), Registration No.: 2009626 (Mex.).
54 Derecho a la Libertad Personal y Derecho a la Privacidad. Su Limitación es Excepcionalísima y Corresponde a la Autoridad Justificar su Afectación, Amparo directo en revisión 3998/2012, 12 November 2014. Thesis published on the Semanario Judicial de la Federación on 13 March 2015.
55 Ley General de Protección de Datos Personales en Posesión de Sujetos Obligados, Diario Oficial de la Federación [D.O.], 26 de Enero de 2017 (Mex.).
56 Derecho a la Intimidad o a la Vida Privada. No se Viola Por el Hecho de Que el Imputado o Algún Miembro de su Familia Sea Fotografiado en la Vía Pú Blica Como Parte de las Labores de Vigilancia e Investigación Para la Perse Cución del Delito Correspondiente, S.C.J.N., Gaceta del Semanario Judicial de la Federación, Décima Época, Tomo II, Libro 20, julio de 2015, Tribunales Colegiados de Circuito, p. 1719, Tesis: II.1o.29 P (10a.), Registro: 2009626 (Mex.). Another limit to Intimacy and Privacy is that of freedom of expression. According to the court, in particular when in the presence of a matter of public interest, information can be diffused or disclosed, that belongs to the personal sphere of individuals [Libertad de Expresión. el Interés Público Constituye Una Causa de Justifica Ción Para Difundir Información Privada, S.C.J.N., Semanario Judicial de la Federación y su Gaceta, Décima Época, Tomo 1, Libro XX, mayo de 2013, Primera Sala, p. 549, Tesis: 1a. CLV/2013 (10a.), Registro: 2003628 (Mex.)].
57 Derecho Fundamental Al Honor. Su Dimensión Subjetiva y Objetiva, S.C.J.N., Semanario Judicial de la Federación y su Gaceta, Décima Época, Tomo I, Libro 3, febrero de 2014, Primera Sala, p. 470, Tesis: 1a./J. 118/2013, Registro: 2005523 (Mex.).
58 However, it should be noted that Canada does have strong commercial data protection laws. See PIPEDA, SC 2000, c 5.
59 Although, under Mexican doctrine it is not entirely clear if "security" is the right term to use in this context. Mexican law protects the possibility to maintain a space in which the individual can develop her personality without interference –

however, judges and legal scholars have not yet clarified whether this refers more to isolation, secrecy, or anonymity.

References

Arteaga Botello, Nelson. 2015. "Surveillance Footage and Space Segregation in Mexico City." *International Sociology* 30(6): 619–636.

Arteaga Botello, Nelson. 2016. "Regulación de la videovigilantcia en México. Gestion de la ciudadanía y acceso a la ciudad." *Espiral: Estudios sobre Estado y Sociedad* 23: 193–238.

CDHDF (Comisión de Derechos Humanos del Distrito Federal) Dirección General de Comunicación por los Derechos Humanos. 2016. *CDHDF documenta tres quejas por presunta violación de DDHH con utilización de Periscope.* Bulletin 031/2016 of February 23, 2016. Mexico City: CDHDF. Accessed June 12, 2017. http://cdhdf.org.mx/2016/02/cdhdf-documenta-tres-quejas-por-presunta-violacion-de-ddhh-con-utilizacion-de-persicope/.

Gutiérrez Ortiz Mena, Alfredo. 2014. "El derecho a la intimidad en la era de la tecnología de las comunicaciones: una reflexión desde el Derecho Constitucional." In *Revista Mexicana de Derecho Constitucional,* N. 31, July–December, pp. 239–245. Mexico City: Instituto de Investigaciones Jurídicas, UNAM.

Haggerty, Kevin D., and Richard V. Ericson. 2006. "The New Politics of Surveillance and Visibility." In *The New Politics of Surveillance and Visibility*, edited by Kevin D. Haggerty and Richard V. Ericson, pp. 3–25. Toronto: University of Toronto Press.

Ibarra Sánchez, Ernesto. 2012. "Videovigilancia. Puntos de colisión entre derechos fundamentales y protección de datos personales en México." In *Jurismática: El Derecho y las Nuevas Tecnologías*, edited by Ernesto Ibarra Sánchez and Rodolfo Romero Flores, pp. 231–269. Monterrey: UANL.

Kerr, Orin S. 2015. "Katz Has Only One Step: The Irrelevance of Subjective Expectations." *University of Chicago Law Review* 82: 113–134(2015).

Lyon, David. 2007. *Surveillance Studies: An Overview.* Cambridge: Polity Press.

Slobogin, Christopher. 2007. *Privacy at Risk: The New Government Surveillance and the Fourth Amendment.* Chicago, IL: University of Chicago Press.

Suprema Corte de Justicia de la Nación. 2016. *Criterios del Poder Judicial de la Federación.* Tercera edición. Ciudad de México: Suprema Corte de Justicia de la Nación.

Téllez Valdés, Julio. 2012. "La regulación jurídica de la videovigilantcia bajo una perspectiva de derecho comparado." In *Jurismática: El Derecho y las Nuevas Tecnologías,* edited by Ernesto Ibarra Sánchez and Rodolfo Romero Flores, pp. 270–285. Monterrey: UANL.

Whitebread, Charles H., and Christopher Slobogin. 2015. *Criminal Procedure: An Analysis of Cases and Concepts.* Mineola, NY: Foundation Press.

Index

Page numbers in **bold** refer to figures, page numbers in *italic* refer to tables.